Praise f

"Dives deep into the investigat[...] researched account informs and [...]"
—*Publishers Weekly*, starred review

"A gripping and brilliantly researched history of the rise and fall of
the Portland spy ring, which reveals much about the operations and
personnel of Russian, British, and American intelligence at the height
of the Cold War."

—Christopher Andrew, author of
The Secret World: A History of Intelligence

"Eye-opening . . . fans of Furst, Ludlum, and their kind will find this
real-world exploration of old-school espionage suitably intriguing."

—*Kirkus Reviews*

"First-rate. . . . Barnes performs an incredible deep dive into this story
of international espionage and counterintelligence, mining both existing sources and newly declassified information to fuel a narrative as
compelling as a spy novel."　　　　　　　　　　　—*Booklist*

"Excellent and riveting, with a cast of characters as engaging as in any
novel. Former KGB officer Vladimir Putin's modern-day Russia em-
ploys the same espionage methods now against the West. The themes
of *Dead Doubles*—deception, betrayal, blackmail, chemical and biological weapons, atomic secrets, international rivalry—are as topical
today as in the 1960s."

—John Sipher, former head of the CIA's Russian operations
and former CIA station chief in Asia and Europe

"An enthralling account of one of the last great spy mysteries of the
twentieth century—I loved it."

—John Preston, author of *A Very English Scandal*

"Gripping . . . reads like a Graham Greene or Le Carré novel. . . .
Anyone interested in Russia's continuing undermining of the West,
espionage, or simply a good thriller read should delve into this book."
—*New York Journal of Books*

DEAD DOUBLES

THE EXTRAORDINARY WORLDWIDE
HUNT FOR ONE OF THE COLD WAR'S
MOST NOTORIOUS SPY RINGS

TREVOR BARNES

HARPER

NEW YORK · LONDON · TORONTO · SYDNEY

HARPER

A hardcover edition of this book was published in 2020 by HarperCollins Publishers.

Originally published in the United Kingdom in 2020 by Weidenfeld & Nicolson, an imprint of the Orion Publishing Group Ltd.

HarperCollins books may be purchased for educational, business, or sales promotional use. For information, please email the Special Markets Department at SPsales@harpercollins.com.

FIRST HARPER PAPERBACKS EDITION PUBLISHED 2021.

Library of Congress Cataloging-in-Publication Data has been applied for.

ISBN 978-0-06-285700-2

21 22 23 24 25 LSC 10 9 8 7 6 5 4 3 2 1

To my wonderful wife, Sally Gaminara

Contents

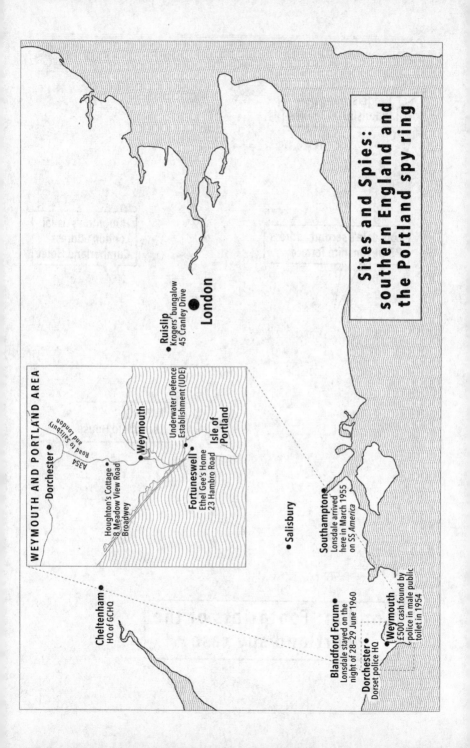

Sites and Spies: southern England and the Portland spy ring

Ruislip
Krogers' bungalow
45 Cranley Drive

London

Cheltenham
HQ of GCHQ

Blandford Forum
Lonsdale stayed on the
night of 28-29 June 1960

Dorchester
Dorset police HQ

Weymouth
£500 cash found by
police in male public
toilet in 1954

Salisbury

Southampton
Lonsdale arrived
here in March 1955
on SS America

WEYMOUTH AND PORTLAND AREA

Dorchester

Houghton's Cottage
8 Meadow View Road
Broadwey

A354
Road to Salisbury
and London

Weymouth

Underwater Defence
Establishment (UDE)

Fortuneswell
Ethel Gee's Home
23 Hambro Road

Isle of Portland

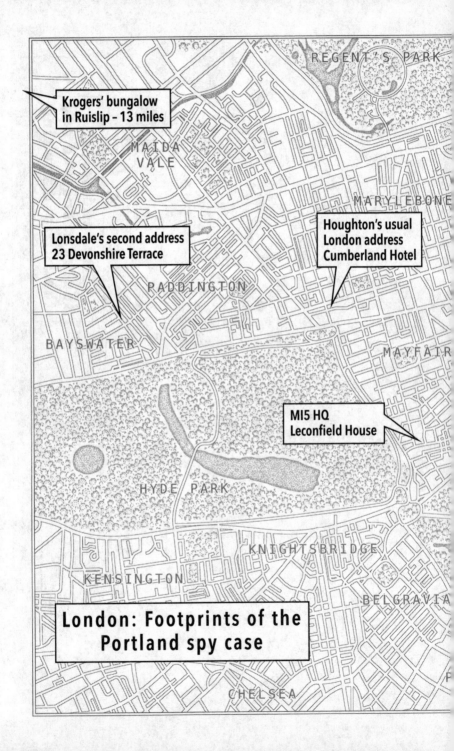

Krogers' bungalow
in Ruislip – 13 miles

REGENT'S PARK

MAIDA
VALE

MARYLEBONE

Lonsdale's second address
23 Devonshire Terrace

Houghton's usual
London address
Cumberland Hotel

PADDINGTON

BAYSWATER

MAYFAIR

MI5 HQ
Leconfield House

HYDE PARK

KNIGHTSBRIDGE

KENSINGTON

BELGRAVIA

London: Footprints of the
Portland spy case

CHELSEA

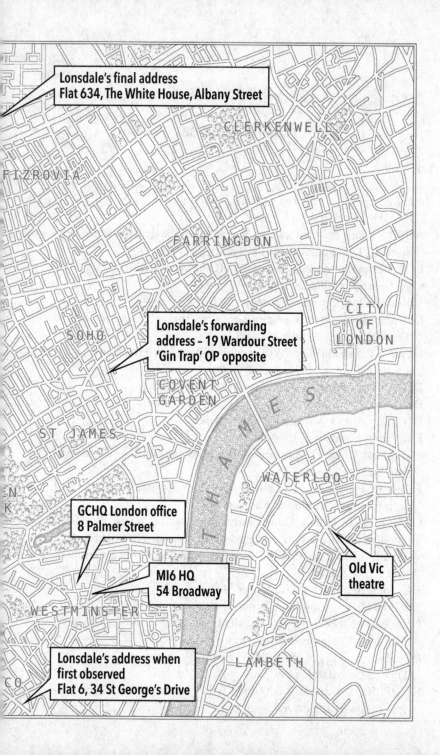

Abbreviations

A2: MI5 technical support section

A4: MI5's 'watchers' department

AUWE: Admiralty Underwater Weapons Establishment (which subsumed UDE in an autumn 1960 reorganisation)

BOB: Berlin Operations Base (CIA HQ in West Berlin)

CIA: Central Intelligence Agency

D Branch: MI5 counter-espionage

D1: MI5 Soviet counter-espionage. Also the abbreviation for the Director of D1, Arthur Martin

D2: MI5 Polish and Czech counter-espionage. Also the abbreviation for the Director of D2, David Whyte

DDNI: Deputy Director of Naval Intelligence (UK)

DG: Director General of MI5

DNI: Director of Naval Intelligence (UK)

FBI: Federal Bureau of Investigation

FCD: First Chief Directorate of the KGB (responsible for foreign intelligence)

FSB: intelligence agency responsible for domestic security and counter-espionage in Russia from 1991

GCHQ: Government Communications Headquarters, the UK's intelligence and security organisation responsible for providing signals intelligence (SIGINT) and information to the government and armed forces

JIC: Joint Intelligence Committee

MI6: the UK's foreign intelligence agency

NSA: National Security Agency, the USA's equivalent of GCHQ

NZSS: New Zealand Security Service

OP: observation post (MI5 abbreviation)

OTP: one-time pad (cipher pad)

RCMP: Royal Canadian Mounted Police (Canada's intelligence as well as law-enforcement agency)

RIS: Russian Intelligence Service (MI5 acronym)

SIS: Secret Intelligence Service – another name for MI6

SLO: security liaison officer (MI5)

SVR: the foreign intelligence service of the Russian Federation from 1991

UB: Urząd Bezpieczeństwa, security service in post-war communist Poland

UDE: Underwater Detection Establishment, Portland

Who's Who and Code-names

Abel, Rudolf: KGB illegal in USA (real name Willie Fisher)

Angleton, James Jesus: head of CIA counter-intelligence

'Asya': KGB code-name for Ethel Gee

Austen, Captain Nigel: British naval attaché in Warsaw 1951–2

Baker, Molly: a London business partner of Lonsdale

Belmont, Alan: head of the FBI's Domestic Intelligence Division from 1951

'Bevision' (or 'Vision'): CIA code-name for agent who provided first tip-off about Houghton

Bonsall, Arthur 'Bill': GCHQ head of Z Division

Bowers, Michael: a London business partner of Lonsdale

Brook, Sir Norman: Cabinet Secretary 1947–62

Butler, Richard Austen ('Rab'): Home Secretary 1957–62

Carrington, Lord Peter: First Lord of the Admiralty

Caswell, John F.: deputy chief of CIA London station

Cohen, Lona: real name of Helen Kroger, code-named 'Mrs Killjoy' (MI5) and 'Dachniki' (KGB)

Cohen, Morris: real name of Peter Kroger, code-named 'Killjoy' (MI5) and 'Dachniki' (KGB)

Colfer, William 'Bill': D1 officer who worked with Elwell on the Lonsdale investigation

Craggs, James: MI5 case officer in D2 (pseudonym at request of Security Service sources)

Cumming, Malcom: MI5 Director of A Branch (General Services)

'Dachniki': KGB code-name for Krogers while based at Ruislip

Denning, Admiral Nigel: Director of Naval Intelligence

Douglas-Home, Alec (Lord Home): UK Prime Minister 1963–4

Dozhdalev, Vasili: KGB officer based in Russian embassy in London

'Dust Cover': MI5 code-name for the OP at the house of Bill and Ruth Search in Ruislip

Elwell, Charles: MI5 officer in counter-espionage D Branch

Feklisov, Alexander: KGB officer in New York

Ferguson Smith, Chief Inspector: Metropolitan Police Special Branch

Fisher, Willie: real name of KGB illegal in USA known as Rudolf Abel

Fuchs, Klaus: German KGB spy in Los Alamos

Furnival Jones, Martin: head of MI5 counter-espionage, D Branch, known as 'D'

Gee, Ethel 'Bunty': clerk in UDE drawing office, girlfriend of Houghton, code-named 'Trellis' (MI5) and 'Asya' (KGB)

Glass, Ann: maiden name of MI5 officer who married Charles Elwell

Gold, Harry: American KGB spy who acted as courier for Klaus Fuchs's atomic secrets

Goleniewski, Michał: important CIA asset in Polish intelligence

Greenglass, David: American KGB spy at Los Alamos (brother of Ethel Rosenberg)

Grist, Evelyn: head of MI5's transcription section, A2A

Hall, Ted: KGB spy at Los Alamos

Hollis, Roger: Director General (DG) of MI5

Hoover, J. Edgar: Director of the FBI

Houghton, Harry: clerk at UDE, code-named 'Reverberate' (MI5) and 'Shah' (KGB)

Johnson, Olive 'Peggy': divorced wife of Harry Houghton

Kennedy, John: President of the USA 1960–63

Kroger, Helen: cover name of Lona Cohen

Kroger, Peter: cover name of Morris Cohen

'Killjoy(s)': MI5 code-name for Peter Kroger/the Krogers

'Last Act': MI5 code-name for Gordon Lonsdale

'Lavinia': MI5 code-name for CIA agent (Goleniewski) who provided tip-off about Houghton

Leggett, George: MI5 officer in D2

Lonsdale, Gordon: KGB illegal working undercover in the UK (real name Konon Molody)

Macmillan, Harold: Prime Minister of the UK 1957–63

Manningham-Buller, Sir Reginald: Attorney General (UK)

Martin, Arthur: MI5, Director of D1 (Soviet counter-espionage), known as 'D1'

Mitchell, Graham: MI5 Deputy Director General (DDG)

Molody, Konon: real identity of Gordon Lonsdale, code-named 'Last Act' (MI5)

Moore, Bridget: MI5 secretary, later wife of David Whyte

Pavlov, Vitali: KGB officer in Illegals Directorate S, deputy chief 1954–9, when appointed chief

Pigot, Tony: Deputy Director of UK Naval Intelligence

'Reverberate': MI5 code-name for Harry Houghton

Roman, Howard: CIA operations officer based in Washington DC

Romer, Sir Charles: chair of the Romer Inquiry, 1962

Rosenberg, Ethel and Julius: American KGB spies during and after the Second World War, who helped to handle the network of KGB spies in the Manhattan Project

Semyonov, Semyon: KGB controller of Cohens in New York 1940–44

'Shah': KGB code-name for Harry Houghton

Shergold, Harold: head of MI6 Sovbloc section, Russian intelligence specialist

Skardon, William 'Jim': head of A4, MI5's watchers

Smith, Superintendent George G.: Metropolitan Police Special Branch

'Sniper': FBI code-name for the CIA agent who provided tip-off about Houghton, real name Michał Goleniewski

Sokolov, Yuri: KGB controller of Cohens in New York, 1947–50

'Trellis': MI5 code-name for Ethel Gee

'Vision' (or 'Bevision'): CIA code-name for its agent who provided tip-off about Houghton

Watford, Alfred: UDE, recipient of anti-Semitic letter

White, Sir Richard 'Dick': head of MI6 (or SIS), known as 'C'

Whyte, David: Director of D2, Polish and Czech counter-espionage at MI5, known as 'D2'

Winterborn, Hugh: MI5, head of A2 (technical support)

Wright, Peter: MI5, worked in A2, recruited as Security Service's first scientist

Yatskov, Anatoli: KGB controller of the Cohens in New York 1944–6

Note on the KGB

KGB is used in this book to mean the Soviet State Security organisation throughout its history from its foundation in 1917 until 1991, as well as specifically the period 1954–91. In 1991 the all-powerful KGB was broken up: its responsibilities for domestic security and counter-espionage within the Russian Federation were given to the FSB, while the SVR took over foreign intelligence. Between 1917 and 1954, Soviet State Security operated under a bewildering number of names after it was first formed as the Cheka, including GPU and GUGB (when incorporated in the NKVD), and then NKGB, MGB and MVD before emerging as the KGB (meaning Committee of State Security) in 1954. Within the KGB proper the section responsible for foreign intelligence was the most prestigious, the First Chief Directorate (or FCD). The department responsible for illegals, S for 'special', was within the FCD.

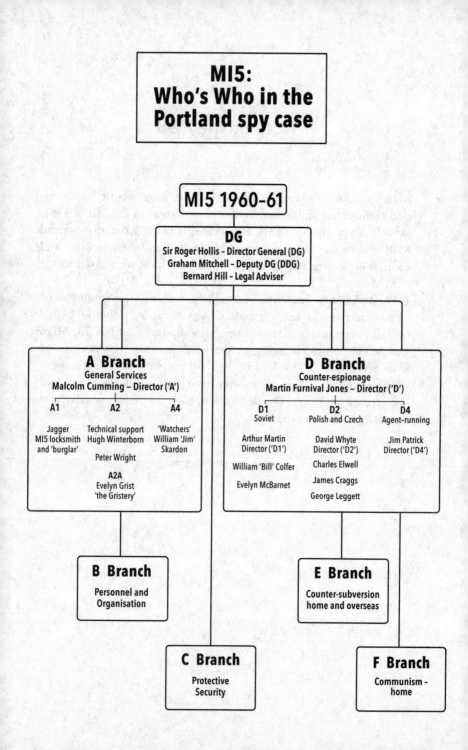

In the higher ranges of Secret Service work, the actual facts of many cases were in every respect equal to the most fantastic inventions of romance and melodrama. Tangle within tangle, plot and counter-plot, ruse and treachery, cross and double-cross, true agent . . . were interwoven in many a texture so intricate as to be incredible yet true.

Winston Churchill[1]

Each mark I've made, my every feature,
Each date's removed as by a hand, erased:
A soul that once was born – somewhere.

So it is my country couldn't keep
Me and the most clever, keenest detective,
Studying my soul, however deep,
Won't find it out – my hidden birthmark.

Marina Tsvetayeva
'Longing for the motherland'[2]

Prologue

London, 12 September 1960

In the early afternoon an unmarked car drove along Great Portland Street on the eastern boundary of Fitzrovia, then still lined with showrooms for the women's ragtrade. The vehicle drew up outside number 159, a branch of the Midland Bank, at the corner with Weymouth Street. The two men inside were from MI5, Britain's counter-espionage service. They glanced up and down the street before they entered the five-storey building shortly before it locked its doors (banks only opened on weekdays and closed at 3.30 p.m. in 1960) and asked for the manager. He and a bank inspector, sent specially from head office, treated the visitors with intrigued deference, knowing that people at the pinnacle of the bank had ordered full cooperation. They unlocked the strongroom and extracted a large paper parcel and an attaché case – items belonging to one of their clients, Gordon Lonsdale. Lonsdale, a Canadian businessman, had recently been placed under observation by the Security Service because he was suspected of being a Soviet spy. Just over a fortnight before, Lonsdale had been seen by MI5's 'watchers' to enter the bank and deposit an attaché case, a briefcase and a deed box, telling the branch that he was leaving shortly for Canada and would return on 26 September. The Security Service could find no trace of Lonsdale departing afterwards from Britain by land, sea or air. He had quite simply vanished.[1]

Lonsdale's absence was providential for MI5. Clearly the items he had deposited at the bank were of value and, if examined covertly, could provide crucial information about his activities. This was a heaven-sent opportunity to discover more about him. It was, however, perilous. No one knew exactly how long Lonsdale might be abroad. Although he said he would be absent for four weeks or so, he might return at any moment, and it was paramount that there should be no indication at all that his belongings had been searched. There was a risk it might all go horribly wrong: Lonsdale might spot some tell-tale clue of MI5's undercover work and instantly be gone.

MI5 were nervous, from the officers and technicians who would carry out the top-secret operation to the very apex of the Service, Roger Hollis, the Director General, who would be in the firing line if it went awry. MI5 had done their utmost to limit the risk. At their behest, the head civil servant of Her Majesty's Treasury had obtained formal approval for the operation from the chairman of the Midland Bank. MI5 had made preparations in the greatest secrecy, including the recruitment of an expert 'remover-putter-back' from the UK's foreign intelligence service, MI6. The operation was of dubious legality, even in the Cold War Britain of 1960, when no intelligence agency officially existed and oversight was largely informal and based on mutual trust.

Carefully guarded by the branch manager and the bank inspector, the items deposited by Lonsdale were driven to the secret MI5 laboratory at St Paul's two miles to the east. Still blinking with amazement as though woken from a dream, the manager and inspector were greeted there by a clutch of Security Service men. The paper parcel was photographed and painstakingly unwrapped. Inside were a brown leather briefcase and a small metal deed box. A towering man called Jagger stepped forward, dressed in a black undertaker's suit and with shoulders as wide as an armchair. A former sergeant-major in the Rifle Brigade, Jagger was the Security Service's factotum. His particular speciality was picking locks. He set to work, caressing the locks with his skeleton keys. For a tall man his hands were surprisingly delicate, and within a few minutes both items were open and their contents laid out on trestle tables under the unforgiving fluorescent lights.

At 3.45 p.m. the door to the laboratory swung open. An elegant man of middle height in a pinstripe suit strode in and hung up his bowler hat. He was Charles Elwell, aged forty-one, the MI5 case officer in charge of the Lonsdale investigation. Elwell nodded to his colleagues, watched Jagger for a few moments labouring to open the attaché case, and turned to scrutinise the documents from the briefcase arranged on a table nearby. Bright desk lamps were positioned. As the unpacking continued notes were scribbled. Cameras clicked incessantly as numerous photographs were taken at all stages of the search to ensure the Security Service possessed a comprehensive record of any discoveries and how they were packed. Most of the documents concerned Lonsdale's business affairs, and especially his bankrupt jukebox business. There was some private correspondence, including love letters, and two books, *Contract Bridge Made Easy* and *Touch Typing in Ten Lessons*. When a narrow beam of light was shone across some of the pages of the books and they were examined under a magnifier there were excited murmurs: indentations were visible, made by someone when writing figures on a piece of paper placed on the page. These were photographed to see if the marks could be deciphered later.

Jagger's gloved hands feathered various keys in the two special five-lever locks of the Skyline attaché case. The locks clicked open. Gingerly, one by one, the contents were unpacked and carefully rested on another table. In a zip-fastened case was a Praktica camera with various lenses and photographic equipment, and separately a cassette holding 100 feet of 35mm film and two magnifiers. There was an address book and sundry other items such as a photographer's black cloth, some letters, more camera film and an Automobile Association membership card. Photostats of the address book and documents were made. There was also a Ronson cigarette lighter fixed in a round wooden base about five inches in diameter. Although some of Lonsdale's belongings were suspicious, MI5 had uncovered no incontrovertible evidence that he was a KGB spy. The disappointment was palpable. One by one they started carefully to repack the items in the attaché case.

Elwell thought it odd that Lonsdale had included the cigarette lighter with the other items. It was possible that he valued it 'for sentimental reasons', as Elwell later wrote in his report, 'though from our knowledge of [Lonsdale] he would not appear to be a sentimentalist'.[2] Elwell suggested the lighter be scrutinised more carefully. It was placed under an X-ray machine, which revealed a mysterious shadow in the base. The metal lighter on the top was unscrewed from the base. A pad of green baize in the hollow beneath hid a screw which, when removed, gave access to a concealed cavity. There were gasps of amazement when, using a rubber suction cup and tweezers, various tiny items were laboriously extracted and placed on the table: three miniature one-time cipher pads (OTPs) and a folded piece of paper listing the names of eight roads in the Kingston area of south-west London and their grid references on a road atlas.

Elwell and his MI5 colleagues had studied USSR secret espionage communications as part of their work. On seeing the cipher pads they were confident that they were of a type used by Soviet intelligence.[3] The OTPs consisted of tiny plastic pages gummed together at the edges in red and black portions, one used for deciphering incoming radio messages and one for encoding outgoing ones. They could only be read or photographed in full by breaking the glue. The head of MI5 counter-espionage, Martin Furnival Jones, was phoned for urgent advice. He ordered that nothing be done to alert Lonsdale that his possessions had been tampered with. Photostat copies were made only of the pages of the OTPs which were already visible and of the Kingston-area list of streets. There had been a number of other objects in the briefcase, the attaché case and the box, including numerous photographs of people and places clearly taken by Lonsdale while in the UK. But none had seemed of immediate interest. The repacking was completed, Jagger successfully locked the attaché case again, the paper parcel was resealed and it was all handed back to the bank manager at 8.30 p.m. The whole operation had taken five hours from start to finish.

The discoveries were stunning. MI5 had uncovered not only a complete set of Russian Cold War espionage paraphernalia, but also proof that Lonsdale was a deep-cover KGB intelligence officer. As he left the laboratory Elwell's excitement was soured by a nagging worry. The smallest of the OTPs extracted from the cavity in the cigarette lighter had been found to be wrapped in a piece of foolscap paper and secured with a rubber band. When first opened, the rubber band had snapped because it was perished. To try to conceal the problem, the technician could only tuck the broken end of the original rubber band under itself. Would Lonsdale notice this when he picked up his possessions? Would anything else suggest his belongings had been tampered with?

Part One
Investigation

1

Code-name 'Reverberate'

I

One notable absentee at the MI5 laboratory on 12 September 1960 was Elwell's boss at the Security Service, David Whyte. Until then it was he who had been heading the investigation involving Lonsdale for seven months, and it was only an event of momentous personal significance that had kept him away. Whyte worked with his friend Elwell in the counter-espionage branch of MI5. He headed section D2, which specialised in Polish and Czech counter-espionage. Back in February it was Whyte who had read a police report that first triggered the investigation.

The report was marked 'Secret' and was from Special Branch in Dorset.[1] The Security Service at this time had a limited number of personnel – only 174 'officers', of whom thirty-five were stationed overseas to liaise with foreign and Commonwealth intelligence organisations. The Security Service relied on Special Branch – police officers in the 150 or so forces stretched across the country outside London – to hoover up and feed back information on potential threats of espionage and subversion of the state.[2]

The report, from a detective constable based in the seaside town of Weymouth, was dated 18 February 1960. It concerned a man working at a highly sensitive naval facility called the Underwater Detection Establishment (UDE) at Portland, near Weymouth. Alfred Watford had been sent by post 'a sheet of foolscap paper on which was drawn in ink a swastika, with the word JEW written underneath'. Watford had complained about this distressing incident to the Admiralty police at the base and told them he suspected a man who also worked at the UDE, Harry Houghton, of sending the letter.

Anti-Semitism was not for the Security Service to investigate. Dorset Special Branch had lodged the report for other reasons. Watford alleged that around 1955 some secret files had disappeared from the strongroom at UDE 'for some days'. According to Watford, Houghton admitted taking them, 'saying he had nothing better to do at the time and that he took the files and read them'. Watford added that Houghton was a former master-at-arms (chief petty officer) in the Royal Navy, had previously worked at the British embassy in Poland and currently 'lives beyond his salary, [and] drinks considerably'. The memo concluded by asking the Security Service for guidance.

The report had been forwarded to Whyte on 26 February by a colleague in the branch of MI5 that kept watch on communists in the UK. Although the allegations about the secret files being borrowed in 1955 were stale, Whyte decided his section was interested – but in a lukewarm way. There was no need to rush out a response.

Compared to the silence and smoke-free air of modern offices, in 1960 the corridors of MI5 resounded with the clacking of Imperial 66 manual typewriters and the air was often fuggy from cigarettes. Outside Whyte's office were seated, as throughout the Security Service, a clutch of women typists. A handful were debutantes, living in Kensington or Bayswater, who treated the whole business with an air of flighty entitlement. The majority were from respectable middle-class families, recruited direct from secretarial college, who dutifully told friends they were employed by the 'War Office' and treated the often dull work with due seriousness but took little interest in its content. At that time, MI5 – like all Western institutions – was deeply sexist. From the late 1950s women had started to be involved in surveillance, but they were not included in the agent-running sections of the Service until 1969.[3] In mid-March, Whyte asked his secretary to come into his room in MI5 headquarters at Leconfield House in Curzon Street, Mayfair. Dressed in a three-piece suit, sporting a bow-tie, and a couple of inches shy of six feet tall, he walked back and forth, as was his custom, while he dictated two notes in his educated baritone. He asked his colleagues in the 'communism – home' branch to extract for him from Admiralty archives details of Houghton's work in Poland, but proposed they should keep charge of the case.

Whyte was a clever and cultivated man with an impish sense of humour. Born in 1915, he graduated from Cambridge in 1936 and served with Special Forces in Yugoslavia and Albania during the war, before being recruited by MI5 in 1947. An early posting for Whyte was to the Soviet counter-espionage section. He was no diehard conservative, reading both the liberal *Guardian* on weekdays and *Observer* on Sundays, as well as the then newspaper of the establishment, *The Times*, every day. He was slightly self-conscious of the 'essential' (uncontrollable) tremor in his right hand but concealed it well, joking that this was why he found it easier to pour out generous measures of spirits. Despite the tremor he was a skilled pianist, and adored listening to opera.[4]

The momentous event which was to compel Whyte to miss the covert search of Lonsdale's possessions in the MI5 laboratory six months later was linked to his love life. He had met an office secretary, Bridget Moore, just before Christmas and in the early months of 1960 their romance was starting to blossom. Bridget was fourteen years younger than Whyte, who had been attracted by her hearty laugh and independent character. They preferred to keep the relationship secret from colleagues at the office: the habit of concealment was now woven into the texture of their lives and if they married strict MI5 rules would compel one of them to leave the service.

Details about Harry Houghton trickled back from Naval Intelligence over the following weeks. Born on 7 June 1905 in Lincoln, he had joined the navy when seventeen, and served continuously until he was demobbed with a pension in December 1945. His character 'was very good throughout'. Houghton was currently employed as a clerk in Portland's Port Auxiliary Unit, where he had some limited access to classified information. More intriguingly for the Security Service, the Admiralty had already exchanged correspondence with MI5's Protective Security Branch about Houghton in the summer of 1956. When that thin file emerged from the bowels of the Security Service Registry, it told an ambiguous story that was to unsettle the most senior management of MI5 in the year ahead. In June 1956, just under five years earlier, the head of the UDE had sent a report about Houghton to the Admiralty. The background to the report was 'domestic strife' between Houghton and his wife, which had caused her to leave him and seek a divorce. During 'recent welfare enquiries', he wrote, Mrs Houghton 'alleged that her husband was divulging secret information to people who ought not to get it. No further action other than discreet surveillance is being taken at this time.' He introduced this information about Houghton by remarking – based on nothing more than the widespread misogynistic views of the time – that 'the whole of these allegations may be nothing more than outpourings of a jealous and disgruntled wife'.

The Admiralty had forwarded this report to MI5 with a covering note, which disclosed that Houghton had been sent home from Poland because he had become very drunk on one occasion and 'it was thought he might break out again and involve himself in trouble with the Poles'. As for Mrs Houghton's allegations about him divulging secret material, the Admiralty commented that 'it seems not unlikely' that they were 'made on the spur of the moment and out of pure spite'. They hesitated to trouble MI5 except for the fact that the department had no record of a basic check being made when Houghton was offered the job at UDE, and that no security information was held on him. This Admiralty letter had landed on the desk of a young MI5 officer. He had checked with the Registry and confirmed that MI5 had no file on Houghton, so he responded to the Admiralty in July 1956: the Security Service 'have no adverse trace of Houghton and agreed that prima facie the allegations seemed to be mainly spiteful. We should be interested to know if you hear of anything to confirm them.' That young officer later had good cause to regret his lazy and biased thinking, which merely echoed the deep-seated prejudices of the Admiralty.

Having reviewed all the correspondence, MI5's 'communism – home' branch considered that 'someone must make a start [on the case] and it might as well be us', setting up a file on Houghton and asking the Dorset police to make further enquiries at Portland.

The next day, a totally unexpected event intervened. At the time, and for years afterwards, it was guarded with the utmost secrecy. It jolted MI5's meandering investigation from an amble to a sprint.

II

The roots of that unforeseen event in April 1960 tangled back to spring 1958, when the US ambassador in Bern received a mysterious letter. It contained two envelopes, one addressed to him, the other to the head of the FBI, J. Edgar Hoover. Written in German and signed

'Heckenschütze', meaning 'sniper' in German, the letters offered secret information to the Americans and set out instructions on how to contact the potential spy. Over the coming months the new source started to send the CIA very valuable information, which was handled with the utmost sensitivity. The CIA baptised its new asset with the code-names 'Vision' or 'Bevision', and the FBI (with less originality) 'Sniper'.[5]

No one knew the identity of the CIA spy. The inference from the available clues (in particular the high quality of the information about Poland) was that he was an officer in the Polish intelligence service, known widely as the Urząd Bezpieczeństwa, or UB. 'Sniper' sent several letters to the CIA in the months that followed. Their contents were a closely guarded secret. There were doubts about the reliability of 'Sniper', especially in Washington, fostered by the CIA's suspicious and calculating head of counter-intelligence, James Jesus Angleton, who feared the new asset might be controlled by the KGB. This shadow continued to hang over the source.

In April 1959 the CIA revealed to the British intelligence services that according to 'Sniper' – in what CIA officers described unsmilingly as 'a horrendous take' – two Soviet agents were operating in Britain: one worked somewhere in the navy, the other in MI6. Two years later the second spy was finally revealed to be George Blake, the highly successful and dangerous KGB agent. The clues to the first were very sketchy, little more than that the agent had worked in Warsaw in the early 1950s. The Security Service had begun an investigation, but it was inconclusive. This 1959 inquiry was one of several into Soviet recruitment of British diplomatic staff in Poland in the 1950s.[6]

In April 1960 the CIA received another letter from 'Sniper' containing crucial new, top-secret material about the identity of the spy in the navy: a summary was hand-delivered to MI5 headquarters in Curzon Street, Mayfair, on the 27th. It was this information that spurred MI5's Portland investigation to a gallop. 'Sniper' reported:

> In about 1951 an employee of the British naval attaché's office in Warsaw was recruited. The employee had access to the secret activities and documents of the attaché ... The name of the

employee was given provisionally by ['Sniper'] as 'Huppkener' or 'Happkener' or 'Huppenkort' or some such. The employee was transferred back to England about the beginning of 1953 and assigned to the Admiralty. Because of his importance, he was taken over by the KGB and continued to work successfully for the KGB in London.

With 'Sniper's letter were two documents. One was a letter dated 25 May 1959 from the Polish Vice-Minister of the Interior to the head of the Polish Military Internal Security Corps marked 'Top Secret of Special Importance'. This attached 'An Index of Documents Acquired in the British Embassy by Means of an Agent Penetration in the Period from January 1952 to November 1952', which consisted of ten typed pages listing a total of ninety-nine items. These varied from the *Handbook of British Naval Intelligence*, which contained 'information on the methods of conducting naval intelligence and on the scope of operation of that intelligence', to details held by the British about sonar on Soviet submarines and desertion from the Polish navy. The covering letter from the Polish vice-minister confirmed that the source of this treasure trove of intelligence had since been transferred from the naval attaché's office in the British embassy in Warsaw and been taken over for 'operational contact' by 'Soviet friends'.

Although there was CIA doubt over the reliability of the new information, in the days that followed David Whyte swung his small team into urgent action. He chose two officers to join him on the case. One was George Leggett, half Polish and a friend, with whom he had worked on Soviet counter-espionage cases in the 1950s.[7] The case officer was James Craggs, a sociable bachelor in his late thirties.[8]

Within a few days, attention focused on the man working at UDE whom the Admiralty had asked MI5 about in 1956: Harry Houghton. On 5 May 1960, Craggs spent the day examining files at the Admiralty. These revealed the dates when Houghton was in Warsaw (30 July 1951 to October 1952) and the identity of the naval attaché while Houghton was there, Captain Nigel Austen. A picture of Houghton's life began to emerge. In December 1951 Austen had cautioned the navy clerk for heavy drinking, and the following May Austen wrote again to say Houghton was still drinking excessively.

Houghton was sent home later that year, and on his return to the UK he was posted to the UDE at Portland.

Whyte cranked up the speed of the Portland investigation, working long hours with Leggett and Craggs. Dorset Special Branch confirmed that Houghton was living at 8 Meadow View Road, Broadwey, near Weymouth. They provided the registration number of Houghton's car and, separately, details of a stash of money discovered in mysterious circumstances in 1954 (£500, found in a male public lavatory near Weymouth pier, worth about £12,500 in today's money). The incongruous place where the money was discovered – the cistern – only heightened MI5's suspicions. The Security Service knew this was a favourite place for the Russian secret service to locate a dead drop.[9]

Whyte wished to start intercepting Houghton's phone calls at home. He knew, however, from his contacts inside the Service that there was already a long waiting list. Unless his request was given special priority, it would not be accepted.

At one end of the mahogany-lined corridor on the sixth floor of MI5's headquarters was the canteen. At the other was an unmarked door. Next to the door was a bell and a metal grille. Only certain officers were allowed to pass into the sanctum beyond, where the recording and transcription of intercepted telephone calls took place. In a large square room, Post Office employees made the recordings and passed the fruits of their labours to Security Service transcribers in an adjacent section. Most of the transcribers were women and their work was overseen by Evelyn Grist. Now elderly, she had worked in the Security Service since before the Second World War and was renowned for her love of hats, necklaces and shawls as well as for her formidable personality. Her small empire was known affectionately as the 'Gristery'.[10] It was here where the problem – a lack of transcribers – lay. Whyte knew what to do. He dictated a memo to MI5's Deputy Director General, requesting the necessary Home Office Warrant and asking for the case to be made an urgent priority. His briefing sheet for Mrs Grist asked the transcribers 'to ascertain . . . whether Houghton is at present in touch with the R.I.S. [Russian intelligence service] or P.I.S. [its Polish equivalent] . . . anything which could be interpreted as a clandestine meeting would naturally be of particular interest'.

Meanwhile, Whyte pressed ahead with another urgent task.

III

Whyte knew from previous investigations how crucial it was not to make an enemy of the government department where an espionage suspect worked. Many – especially large and powerful ones like the Admiralty – regarded the Security Service with suspicion, crammed with interfering policemen. The first reaction to news of an MI5 inquiry was often incredulity tinged with hostility. By 1960 Britain's Royal Navy was no longer the behemoth it had been in 1939, but it remained the third-largest in the world after the navies of the USA and the Soviet Union, with nine aircraft carriers, eight cruisers, a startling 114 destroyers, frigates and escorts, and forty-eight submarines.[11] A naval force of this heft had a government department to match, the Admiralty, anchored in grandiose headquarters in Admiralty Buildings in Whitehall, and still with its own separate Cabinet minister, the First Lord of the Admiralty (the post was finally abolished in 1964). Although at times the Royal Navy could be startlingly unconventional, in the early 1960s it was still preserved in the aspic of tradition: at the 1961 Royal Tournament, the navy's displays were of cutlass fighting and hornpipe dancing from the era of Nelson, and of fieldguns from the time of the Boer War.[12]

On 12 May David Whyte, accompanied by Harold Shergold (an officer from MI6 specialising in Soviet and Polish affairs), visited the iconic red-brick Admiralty buildings. From his extensive experience of counter-espionage investigations involving Eastern Europe and the Soviet Union, Whyte knew it was essential to run down every potential clue to the background of a suspect, and he had cultivated contacts in MI6 to help him with this task. Shergold was especially valued because he was regarded as MI6's best Soviet specialist. An alumnus of both Oxford and Cambridge, he had served with Military Intelligence during the Second World War before joining MI6, and was to prove the lynchpin of the investigation in 1961 which finally exposed George Blake as a KGB agent.[13] Shergold was deeply involved in assessing the intelligence provided by 'Sniper'.

Whyte and Shergold made their way to Naval Intelligence. Whyte got straight to the point: Houghton was 'a prime suspect', MI5 were investigating and wanted to send an officer down to Portland urgently. The intelligence officers emphasised the need to check the records of other possible suspects for the leaks of the secret information in Warsaw, and asked to see the welfare officer in Portland who had sent in the report of Mrs Houghton's allegations against her husband. There was a problem here, the Admiralty indicated. The welfare officer was not 'a suitable person ... to have dealings with' because she 'had been pensioned off three years ago for chronic alcoholism'.

Meanwhile, there was a snag over the telephone tap on Houghton: the engineer in Weymouth who was to install it had the same surname as the suspect, Houghton. Whyte asked for some urgent and discreet enquiries to be made locally: the risk was too great if by coincidence the sixty-year-old man was a relative. A few days later, the Post Office rang to say checks confirmed that the engineer in Weymouth called Houghton had no close relatives in the area. The telephone intercepts would start directly.

Step by step, phone call by phone call, memo by memo, Whyte made progress. More information on Harry Houghton percolated through from Dorset police. A young and capable Special Branch detective constable had been assigned to the case. He had an informant whose house overlooked the navy clerk's cottage. This neighbour confirmed that Houghton 'had plenty of money and carried large sums in his wallet'; used to be absent at weekends (Mrs Houghton telling the neighbour that her husband was with the Polish embassy in London); drank 'considerably' and was seen on Easter Monday 1960 in a pub called the Elm Tree near Weymouth 'talking to a man and a woman described as of foreign extraction ... The meeting was obviously prearranged.'

Whyte worked feverishly with MI6 to investigate the list of ninety-nine documents 'Sniper' had provided dating from 1952. British intelligence knew that, having passed on such startling information about the KGB agent in the British naval attaché's office in Warsaw in the early 1950s, the CIA would wish to be briefed urgently on how alarming this breach of security was and its implications.[14] On 12 May 1960, Whyte and Shergold delivered by hand a letter marked 'Top Secret and Strictly Personal' to their contact at the CIA's London office, deputy station chief John F. Caswell, based at 71 Grosvenor Street in Mayfair.[15] This painstaking missive outlined the 'tentative conclusions' of the joint MI6/MI5 investigation into the documents. The two most important were that the British were dealing with separate leaks of sensitive intelligence in Poland, one in 1951–2 and another in 1954–5; but, as regards the 1951–2 documents, MI5 enquiries had 'thrown up one suspect' – Harry Houghton.

Special Branch in Dorset continued to investigate Houghton discreetly. By now he had been accorded the code-name 'Reverberate'. They confirmed that Houghton's ex-wife, after her divorce in 1957, had married a Weymouth man called Herbert Johnson, who was in the Royal Air Force and currently stationed with her in Malaya. According to the police, Johnson had made a 'name for himself in the district as a long-distance walker', and won the *Malay Mail* walking race. So on 23 May Leggett sent a secret telegram to MI5's security liaison officer in Malaya, asking him to locate Johnson. Two days later he confirmed that Johnson was serving with 52 Squadron RAF Kuala Lumpur, and was 'Identical walking race winner'.

Towards the end of May, George Leggett and James Craggs took the train to Dorchester to spend three days in the Portland area. The Britain of 1960 they crossed to reach the county of Dorset was in transition. This was three years before public spending cuts forced the closure of many much-cherished but financially struggling railway tracks, and numerous branch-lines still infiltrated every nook and cranny of the nation. The Prime Minister himself, Harold Macmillan, had opened the first eight-mile stretch of motorway at Preston in Lancashire two years before, but cars were only driving along ninety-five miles of it in total by the end of 1960, compared to 355 miles five years later. And remarkably, during the first decade of the National Health Service after 1948 not a single new hospital was built. It was only from 1962 that a burst of spending was to transform the NHS, spearheaded by a politician whose name is now for many in Britain only a symbol of racism and division: Enoch Powell.[16] Few if any signs of the profound moral and cultural revolution that was supposed to mark the new decade were yet evident. Conformity, continuity and conservatism were the watchwords. As Dominic Sandbrook, a historian of the early 1960s, noted: 'British literature still largely clung to its traditional emphasis on the pragmatic and the idiosyncratic; music-hall traditions endured in songs and television shows; men devoted their weekends to gardening, bowling and fishing; and families decamped on holiday to Scarborough and Skegness.'[17]

This conservatism was especially strong in isolated and self-contained communities like the Isle of Portland. The isle is in fact a peninsula, four miles long by 1.7 miles wide, jutting out into the English Channel. Thomas Hardy described it as 'carved by Time out of a single stone'. A natural haven for shipping, Portland was the site of the first recorded Viking landing on the British Isles in 789 AD, and an informal anchorage for ships in the royal fleet from medieval times until the Royal Navy officially established a major base there in 1845. Portland harbour stretched out its Victorian stone arms across the enormous natural anchorage to the east of the narrow isthmus of Chesil Beach, which joins the mainland to the Isle of Portland. Along that ribbon of land extended not only a carriageway of tarmac but also a railway; a stream of both passenger and goods trains chugged to and fro. Close to the railway track as it approached the isle, several enormous, bulbous grey tanks loomed up into the sky – a key part of the UK's national oil reserve – while navy helicopters chattered across the clouds from the base nearby. The port bustled with activity both day and night. Morse signals flickered across the harbour after dark. During the day, vessels churned the water as they moved from berth to berth, and so many were squeezed together at anchor that visitors had the impression that they could walk across their decks from one side of the port to the other. The centre of the town of Portland was not sleepy and down at heel as it is today, with the usual parade of closed businesses and charity shops lining the main street, but a thriving commercial hub. Nowadays the isle does not have a single bank; back then there were branches of all the major lenders. It had for centuries maintained an aura of being separate and apart, with its own court for various arcane legal matters and its own customs, always a destination and never a thoroughfare.

In the morning of 26 May Leggett and Craggs were driven through the security entrance next to the port and round to the jumble of buildings which constituted the Underwater Detection Establishment. There they met UDE's security officer and interviewed the man, Alfred Watford, whose receipt of the anti-Semitic 'swastika letter' in January had first triggered interest in Houghton. More facts about Houghton's life came into focus: his income (£741 and a naval pension of £160 a year – worth about £20,000 today); his access to classified information (since he had been moved to a new job in January 1957 this was very limited indeed, but previously he had 'direct access to . . . information up to and including Secret');

and that Houghton had been suspected of unauthorised borrowing of files relating to American aid for certain UDE projects, and that this was the reason why his job was changed at the end of 1956.

Houghton had a girlfriend, Ethel Gee, known as 'Bunty', who also worked at UDE. She had been vetted for security, worked as a record keeper in a drawing office at the Portland base, and was a frequent visitor to, but did not live at, Houghton's cottage. The UDE people patronisingly told the Security Service that Gee had access to classified information, but 'for a person of her limited education it would be difficult, if not impossible, to extract information of any value'.

Houghton had a faint North Country accent and, according to some who worked with him, was a little man with rather a lot to say for himself, sometimes in crude language. He had the unpleasant personal habit of keeping his false teeth in his pocket and was too mean to contribute to office collections.[18] Houghton was living at a higher level than other people on his pay grade, buying various new cars and paying builders 'in cash from a large wad of ready money which he always seemed to carry' for major improvements to his cottage. In his early days at UDE he visited London every month, appearing in the office before he left 'looking like a bookmaker'.

Leggett and Craggs were introduced to Houghton's neighbour and Dorset police informant, Cyril Boggust. He was a civilian volunteer special sergeant in the force: 'an ardent special . . . [officer who] would probably have been a regular member of the police force were it not for the fact he has a cast in one eye'. In his early fifties, wiry and energetic, Boggust was married and lived at 5 Rose Cottages, from whose windows he could spy on the front rooms of Houghton's cottage. He struck the Security Service as 'a very stolid, level-headed sort of person'. Boggust told them with chilling nonchalance that during the break-up of his marriage Houghton 'was in the habit of beating his wife from time to time'. It was Boggust's judgement – again, no doubt reflecting male prejudices of the time – that she was 'probably more to blame than her husband for the break-up . . . for he thought she was something of a shrew and . . . was associating with her present husband before 'Reverberate' took a fancy to Miss Gee'. The ex-Mrs Houghton had told Boggust that her former husband was in the habit of visiting the Polish embassy when he was in London and that he 'will get had and he ought to be for what he is doing'. Boggust agreed to keep a watch on Houghton's movements for MI5.

George Leggett conducted follow-up interviews. The most important was with Houghton's former boss in the early 1950s, the British naval attaché in Warsaw, Captain Nigel Austen. It was there, after all, that 'Sniper' suggested the KGB spy in the Admiralty had been recruited.

IV

The rendezvous with Nigel Austen took place at the United Services Club in Pall Mall. There, at the civilised hour of 11 a.m., Leggett and Shergold met the man who had supervised Houghton for a year from August 1951. Before his retirement from the navy in January 1960, Austen had been Deputy Director of Naval Intelligence, so he knew the form. He was forthright: as clerk to the naval attaché Houghton 'was fairly incompetent and typed with his toes judging by the result'. He found his personality 'unattractive', and when he noticed Houghton was drinking heavily he 'lost no opportunity in getting rid of him'. Austen was equally cutting about Houghton's wife, 'a colourless, drab individual who disliked being in Warsaw and no doubt was partly responsible for Houghton's conduct'.

Houghton had full access to all Austen's classified material and could 'quite easily have taken papers from the embassy for quite a long period without their disappearance being noted'. Houghton undoubtedly made money on the black market, 'but probably no more than any other junior employee at the embassy', and when embassy staff returned to the UK they were allowed to sell the bulk of their belongings to local Poles 'at a very handsome profit'. With this cash, Austen estimated, Houghton could well have arrived back in Britain with £2,000 (around £45,000 in today's money) or more. Leggett paused and glanced around the private room at the club before asking his final question: did Austen think Houghton was a spy? Austen replied that although he had a 'fairly low opinion of Houghton, he doubted if he would have gone to the length of selling secrets to the Poles; at any rate there was nothing to indicate this during his time in Warsaw'.

As Whyte and Leggett watched Houghton's buff-coloured MI5

personal file thicken with information, they understood only too well that although there was evidence that Houghton had spied for the Poles and Russians while in Warsaw and perhaps at UDE until the end of 1956, there was none that he was an active KGB agent now. He had had no access to classified information for over three years, if UDE was to be believed, and there was no evidence of any contact with any known Soviet intelligence officers. Whyte and Leggett decided to tighten the surveillance net around Houghton. On 1 June Leggett phoned the Dorset police and agreed that he and William (known as Jim) Skardon would drive down to Dorset to discuss next steps. Skardon, known for being the interrogator who had gained the confidence of and finally cracked the British atomic spy Klaus Fuchs in 1950, had been head of the team of MI5 watchers since 1953.

On 9 June, accompanied by the Special Branch constable, Leggett and Skardon drove south from Dorchester towards Weymouth. At Broadwey, a few miles north of the coast, they forked left. They passed under a bridge carrying the railway line down to Weymouth, parked the car discreetly past the entrance to Meadow View Road and then walked back to reconnoitre on foot. Skardon sketched a crude map of the area. Meadow View Road was a cul-de-sac. On the right were Rose Cottages, a row of five, built from the local grey stone, in the last of which lived Special Sergeant Cyril Boggust. At the end of the cottages was a patch of scrubby garden and a small house; to the right of this was 8 Meadow View Road, and to the left a ramshackle garage where Houghton sometimes housed his car. Boggust's garden and some of his windows looked onto Houghton's whitewashed two-storey cottage, with its small windows peeping from under the black-tiled roof. To the right was a shed in Houghton's untidy back garden, a rusty wire fence, and beyond that rolling open fields where cows grazed peacefully.

Afterwards the MI5 officers visited the naval base to make arrangements to tap Houghton's work telephone extension. The UDE, however, pointed out that he was free to make any calls he wished from the public telephone boxes inside the base. As Leggett strolled past the one nearest to Houghton's office, he saw with a faint shock of recognition the suspect's Renault 'parked within two yards', and noted that 'it may be of significance' that lying on the back seat was the *Geographia Atlas of Greater London*. Special Branch picked up

Cyril Boggust and ferried him to Weymouth police headquarters. Boggust told them that Houghton had been away the previous weekend, leaving home early on Saturday, 4 June, and returning in the afternoon of the following Tuesday with Ethel Gee. He had, of course, no idea where Houghton had gone. Boggust agreed to compile for the police a log of Houghton's movements to and from home, and to raise the alarm if he noticed him obviously leaving for the weekend. In common with many people in the Britain of 1960, Boggust did not have a telephone at home, so he would walk or cycle to the nearest call box a few minutes away to phone the police in Weymouth. MI5's newly recruited civilian agent provided physical descriptions of Houghton ('looks sixty; height 5ft 6 and half/7 ins; grey hair, balding on top; does not wear spectacles; fairly stocky build; dressed respectably; wears grey trilby hat') and Gee ('aged about forty-five to fifty; height 5ft 7/8 ins; well built; upright carriage; brown hair; quite a smart woman').

On 20 June Whyte despatched a 'Top Secret' message to Brian Wise, MI5's security liaison officer in Malaya. Based in Kuala Lumpur, Wise was part of Security Intelligence Far East, a joint MI5/MI6 organisation. Whyte asked him to interview the former Mrs Houghton urgently, stressing the paramount need for secrecy: 'on no account must she say anything about it to anyone, least of all Houghton'.

At the same time MI5 covertly started to open all mail sent to and from Houghton's home address. These checks would take place in the carefully guarded Special Investigations Unit Room in the nearest major sorting office to Weymouth. Here, technicians equipped with large kettles to steam open the mail, rubber gloves to avoid leaving fingerprints, and bright lights, sat at trestle tables opening letters and copying their contents with pedal-operated cameras.[19] Taps by now had been installed on Houghton's telephones at home and work.

On the evening of Friday, 1 July, Houghton telephoned an acquaintance in Manchester. During some general chatter he revealed that he was still 'together' with his 'little girl', clearly referring to his girlfriend, Ethel Gee, and that they had spent the weekend in Eastbourne, a sedate resort on the south coast. Mrs Grist's transcriber then noted Houghton say 'they' were staying the weekend of 9 and 10 July in London: 'We got tickets for the Bolshoi Ballet at the Albert Hall . . . a friend of ours in London she got us some very good seats actually, so think of us next Saturday at the Bolshoi balleying ourselves.'[20]

A few days later Dorset Special Branch sent their first log of Houghton's movements to the Security Service, largely based on information from Boggust. It confirmed that Bunty Gee was a frequent visitor to Houghton's home but never stayed the night, and that he had few visitors. Only one was to prove of interest to the Security Service: on 28 June at 6 p.m. Houghton returned from work, and half an hour later a man was seen to call on him: 'carrying a green holdall-type bag. The man was about 5ft 8ins/5ft 9, wearing dark suit. He was not seen to leave but . . . he had no car.' It was only months later, however, that MI5 deciphered the significance of this visit. On 7 July MI5 intercepted a reservation sent to Houghton from the Cumberland Hotel near Marble Arch in London for a double room for the night of Saturday, 9 July. With clear evidence now from several sources that Houghton was planning a visit to the capital with Ethel Gee that weekend, Whyte made arrangements with Skardon's team of watchers to keep them under surveillance.

On the Friday Whyte received a telegram from Brian Wise in Malaya with pivotal news of the interview with Houghton's former wife. Whyte read the decoded telegram with a mixture of relief and excitement: '1. Husband and wife interviewed yesterday. Chances of blow back seem most unlikely. 2. Details follow by bag. Strong presumption of espionage on part of former husband.'

On the Saturday morning Dorset police saw Houghton pick up Ethel Gee and drive towards London. Skardon deployed his A4 watchers on the main road into London from the south-west to pick up Houghton on the outskirts. Houghton did not appear. There was growing unease in A4 as the hours ticked by and Houghton's car, XOW 513, had not been sighted. At 2.05 p.m., however, the tension eased when the watchers learnt that Houghton and Gee had booked into room 632 at the Cumberland Hotel near Marble Arch at 1.20 p.m. Skardon rushed his officers there. It was cloudy and showery, and at the Cumberland they followed Houghton when he left the hotel at 3.20 p.m. to pick up a plastic mac from his car. He walked out ten minutes later with Gee to travel by Underground from Marble Arch to Waterloo, Skardon noting in his report: 'Miss Gee showing some disinclination to use the escalator, and to

'Reverberate's evident amusement walked down the stairs rather than do so.' Once at Waterloo station, Houghton and Gee took the staircase down to Waterloo Road on the north-east side. One of the watchers tailing Houghton was close enough to hear him say to Gee, 'This is the way we came last time.' They emerged on Waterloo Road and turned right. At the junction with The Cut, opposite the imposing bulk of the Old Vic theatre, was a small public garden, recently opened by Lambeth Council on the derelict site of a building destroyed by bombs during the war.[21]

There they were met at 4 p.m. by a stocky man in his late thirties with dark hair whom A4 identified at that time as a Polish intelligence officer based in London. The trio shook hands and were 'obviously all well known to each other', according to the surveillance officers. They walked east along The Cut. Pockmarked by patches of weed-strewn land, still left waste after the war, the area echoed to the rumble of trains and screech of rails as carriages passed to and fro across the Victorian viaducts which criss-crossed above the roads. The sky was wide and open, not blocked then as it was to be decades later by skyscrapers and towering office blocks. Turning north and later west, past the smoke-blackened terraced houses of Roupell Street, the three suspects ambled back towards Waterloo station. During this walk one of the watchers saw the Polish intelligence officer take what appeared to be two tickets from his wallet and hand them to Houghton, presuming these to be for the Bolshoi Ballet that night. As they passed Waterloo at around 4.30 p.m., Houghton slipped into the station while Gee and the Polish man wandered back to the small park opposite the Old Vic. Here Gee and the new suspect sat down on a bench and chatted for about ten minutes, with Gee 'talking twenty to the dozen', one of the watchers later said, before Houghton reappeared with a blue-paper carrier bag which the surveillance team assumed Houghton had collected from the station. Inside the bag was a parcel about twelve inches by nine by four, wrapped in brown paper. Houghton immediately handed the bag and parcel to the Pole.

All three remained on the bench conversing, 'completely at ease', and while 'they were together it was noticed that Miss Gee did a very great deal more talking than 'Reverberate''. At about 5 p.m. they separated, with the Polish man going off in the opposite direction to the couple, before, 'very alert and on the lookout for followers', walking a 'figure of eight' around the local area 'to discover whether he was being followed'. He then drove off in a grey Standard Estate car parked within 100 yards of the rendezvous. The observation continued but revealed nothing of interest, except that Houghton and Gee took a bus to the Royal Albert Hall for the ballet in the evening. The car driven off by the man presumed to be a Polish intelligence officer was registered in the name of a different individual called Gordon Lonsdale, who lived in a flat in Pimlico.

The next morning Houghton and Gee drove back to Dorset. Having written up his report on the weekend's events on the assumption that the man Houghton and Gee met was a known Polish spy, the ever-prudent Jim Skardon ordered an observation that Monday of the Polish consulate so that the watchers concerned could confirm his identity. In a memo the same day, Skardon informed Whyte: 'At 16.10 Mieczyslaw Kowalski left the Polish consulate and our officers were able to see that, although he bears a strong resemblance to the man Houghton met in Waterloo Road he is, in fact, not identical with that person.' The watchers had mistaken the mysterious man Houghton and Gee had met. The only clue to his real identity was the name and address of the car's owner, Gordon Lonsdale. His address – Flat 6, 34 St George's Drive, SW1 – was immediately put under surveillance. Observation all the next day yielded nothing. The person who had met Houghton and Gee on Saturday was not seen there, and nor was Lonsdale's car. The man had disappeared.

2

Code-name 'Last Act'

I

The 'Reverberate' team did not have a photograph of the mysterious man who met Houghton and Gee, only a description. Jim Skardon painted him as: 'Looks thirty-five. 5ft 9ins. Medium build. Black hair, loosely waved, oval face, pale complexion, pugilistic features, regular teeth, wide nostrils, East-European nationality. Dressed in light-grey sports jacket, centre vent, dark-grey trousers, brown leather shoes, carried blue zipper document case, spoke English with accent.' Meanwhile the focus moved to Gordon Lonsdale. There was no sign of Lonsdale at his flat in Pimlico. It emerged that when he had applied for his British driving licence he had relied on a Canadian one issued in British Columbia. Whyte decided to exploit the Security Service's contacts with the Royal Canadian Mounted Police (RCMP). Canada had never established a foreign intelligence service akin to the CIA or MI6, but by 1960 the RCMP did possess a counter-espionage department, which MI5 contacted through its liaison officer in Canada House in Trafalgar Square. In mid-July Whyte asked the RCMP and also the Metropolitan Police for general background on Lonsdale. Within a week they had established that he was born in Ontario on 27 August 1924, held a Canadian passport issued in Ottawa in January 1955, and had permanently left the Pimlico flat. No one knew his new address. Lonsdale had arrived from Canada in 1955 as a student but by 1960 was describing himself as a company director.[22]

Whyte's attention swung to the eight closely typed pages of the statement marked 'Top Secret' and made by the former Mrs Houghton to Brian Wise in Malaya. Wise had met her second husband, Herbert Johnson, first, before approaching his real object of interest, Mrs Amy Olive Johnson. Although an RAF officer had told Wise that Herbert Johnson was 'pretty dumb, and [you] would need a pickaxe to prise anything out of his head', Johnson attributed his poor memory to 'rough treatment' when held as a prisoner of war by the Japanese. Johnson said 'his wife had had a wretched life with Houghton, who had bullied her for years and had frequently assaulted her'. She no longer had any contact with him, nor did their daughter. Wise met Mrs Johnson (known as 'Peggy') in the family home on the outskirts of Kuala Lumpur. She seemed 'in a very nervous state', and when Wise explained that he wished to ask her a few questions about her former husband, she shot back, 'Oh, what on earth has he been up to now?'

Brian Wise's report on Mrs Johnson was vivid and direct:

> She is a woman of medium height with a great mop of dyed hair, a long nose and rolling eyes. She gives the impression of being extremely highly strung; she talks at length and with great rapidity. It was difficult to pin her down and she constantly flew off at tangents. The description of her as drab and ineffectual is not unfair, and I have no doubt that any spirit or character she ever had was long ago beaten out of her by Houghton. She described herself as simple, and in her own words, a rather stupid person, but . . . she is by no means a nitwit. She has, however, an appalling memory when it comes to dates.

Bearing in mind the way in which her allegations against her former husband, of both possible spying and domestic abuse, had been belittled by the Admiralty years before, Peggy Johnson – remarkably – did not display any bitterness or resentment. She was clearly pleased at last to have a sympathetic listener. 'I think I know what you are after,' she began, 'funny business with Poles.' She said she had remained silent until now through a combination of misplaced loyalty to Houghton and terror of him: he had assaulted her, and threatened 'on numerous occasions to kill her if she opened her mouth about his activities'. Mrs Johnson, evidently very distressed, recited an appalling catalogue of abuse: how, for example, Houghton had broken her leg during one violent fit in Warsaw, injured her by throwing her over a wall, threatened her with a pair of revolvers, and once even attempted to push her off the cliffs at Portland. Wise, evidently no stranger to compassion, noted at this point that her current 'husband made encouraging noises . . . they seemed happily married. Each one, I suspect, gives the other a good deal of comfort and support.'

According to Mrs Johnson, while in Warsaw Houghton was 'frequently the worse for drink in public, and apt to talk loudly and indiscreetly about his work. On . . . occasions, at official parties at the embassy, Captain Austen was obliged to send Houghton home by car, he having become incapable of standing up.' These drunken episodes engendered cruel bouts of wife-beating. Houghton engaged in various activities which puzzled his wife: talking on the telephone between 11 p.m. and midnight in another room to someone he called 'Roger', and going out on Wednesday nights between 9 and 11 p.m. and returning with bundles of banknotes. When quizzed by his wife, Houghton once said 'he was going to string along for the side which paid him the most money', and on another occasion he took the line that he was selling antibiotics on the black market.

Back in Weymouth in 1952, when Houghton was based at the UDE, he had bought and modernised his cottage. According to his ex-wife the drinking bouts continued, as did his abuse and threats. Houghton started driving up to London once a month on a Saturday. When asked about these visits, Houghton replied that 'he had contacts', whom he met at a pub called the Toby Jug on the outskirts of London on the main A3 Portsmouth Road. After returning late on a Saturday night from one of these trips, Houghton, 'having reached the "merry" stage in his alcoholic progress, pulled a bundle of pound notes out of his pocket and shouted "Whoopee" or something similar, and threw them up the air'. Peggy Johnson recalled that he kept a piece of chalk in the glove compartment of his car. His explanation was that 'he required it to make signs in places'.

When he visited London Houghton always took a briefcase with him, which he locked the night before either in the boot of his car or in his desk. On one occasion he brought home a parcel and, after he 'drank himself into oblivion, and while he was snoring away on the couch,' Peggy Johnson untied it and discovered inside a bundle of papers marked 'Top Secret' which she thought concerned torpedoes and underwater detection equipment. Another time, when he was deep asleep following a drinking bout, a paper slipped out of his pocket on to the floor and his wife copied it. She stored it away and gave the MI5 man her copy. Their doomed marriage tottered to its ignoble conclusion: Houghton moved out in September 1955 and lived in a caravan; he cut off Peggy's allowance so she was forced to work as a nurse; Houghton locked her out of the cottage; Peggy sought help from the probation officer in Weymouth (who dealt with personal welfare and legal matters at the base), who suggested she seek legal advice and a divorce, which was finally granted on grounds of cruelty. Shortly afterwards she married Herbert Johnson.

Asked about Ethel Gee, Peggy Johnson remarked that she could not believe 'her former husband's relations with Miss Gee were of a sexual nature', and suggested it was based on Gee's access to classified information. When she reprimanded Houghton for openly flaunting his friendship with Gee, he replied, 'She is useful to me.' Wise concluded his report by saying that 'if Mrs Johnson concocted this story she is a far more wily and intelligent person than I took her to be during our short acquaintance,' and that she was still terrified of Houghton: she said 'that if she came face to face with him in court she would probably faint – and indeed she probably would'.

From the statement, and Houghton and Gee's shifty meeting on 9 July with the still unidentified man, Whyte drew two important conclusions: '"Reverberate" is still actively engaged in spying; and ... Miss Gee is also spying, having probably been recruited by "Reverberate" as a source when he himself was denied access to classified information as a result of his transfer from UDE to Portland dockyard.'

On the evening of Saturday, 23 July, the Security Service eavesdropped on Houghton receiving a telephone call from an unknown woman, Miss Hope Fripp, who spoke with what sounded like an Australian or New Zealand accent. The caller said she had received Houghton's letter and asked how she would recognise him. Houghton called her back after midnight and apologised for phoning and waking her so late, but said that earlier his house was 'full of people' and he 'had not been able to speak as he would have liked to'. He went on: 'On *Saturday week* [the MI5 transcriber accentuated these words and noted in the margin, 'Saturday, 6 August'], then, I will see you, and I'll write to you and describe how I'm dressed and all the rest of it and you'll find me, all right?'

The intercept suggested Houghton was coming to London to meet Miss Hope Fripp on 6 August. Was it connected with espionage? MI5 quickly established that she rented a room in South Hampstead, and set about delving into her background. Was Hope Fripp perhaps a contact of the mysterious Lonsdale? The 'Reverberate' team also started to make some enquiries about two other women Houghton was friendly with. One, an attractive widow called Kath Small, lived in Weymouth, while the other, a Miss Elphick, made an intercepted phone call to Houghton on 15 July, thanking him for a night out and saying she would ring him on her next visit.

The investigation was now so sensitive that Whyte decided he should visit Portland himself. On Tuesday, 26 July, he travelled down to Weymouth with James Craggs. As railway porters in claret and mustard uniforms bustled onto the platform to carry holidaymakers' luggage, the Security Service men were met by a navy car which drove them to Portland dockyard. Seagulls wheeled and cawed in the sky. The roads approaching the isle were stained white by the dust from the lorries groaning up and down laden with Portland stone. On the isle itself the landscape was both gentle and undulating, with grass slopes falling down to the sea, and harsh and angular, where buildings constructed by military bureaucrats threw up graceless naval accommodation, or pockmarked by quarries. Once over the isthmus, with Chesil Beach on the right blocking the view to the glittering sea, the Isle of Portland loomed up ahead against the summer sky. Before climbing up onto the isle, the road swung left into the dockyard area, and beyond it the entrance to UDE, a ramshackle collection of office blocks and laboratories behind which the isle was an ever-present shadow.

In a spartan meeting room, Whyte and Craggs met two men from UDE: the head of administration and Crewe-Read, a retired naval commander who was security officer. Whyte was determined to discover exactly what access Ethel Gee had to secret material and the danger she posed. He found Crewe-Read's answers unclear and unconvincing and eventually the chief of the UDE drawing office was summoned. Gee worked in the registry, which held drawings concerned with projects at the production stage with security classifications up to 'Secret'. Gee was described as 'a good worker and intelligent and would probably be able to appreciate what drawings were important'. (Whyte commented acidly at this point in his note of the meeting: 'This is exactly contrary to what we had been told by ... Crewe-Read.') Gee could also freely visit the drawing office, which housed drawings of projects still in the experimental or prototype stage. The head of the drawing office said: 'it would be perfectly possible for her to take drawings home with her and return them without being noticed. She usually brought a shopping bag to the office.'

Crewe-Read had not yet even informed the new overall head of UDE about the case. Whyte, no doubt biting his tongue, told him he thought he should do so. The final point was to confirm where Ethel Gee lived. The answer was 23 Hambro Road, a brisk walk from the entrance to the dockyard up the hill in an area of terraced streets known as Fortuneswell, built in the early 1900s for workers in the dockyard. Ethel Gee lived a circumscribed life there with her eighty-year-old mother, with whom, remarkably for a forty-six-year-old woman, she still shared a bedroom, her seventy-six-year-old uncle and disabled aunt. It was a two-storey terraced house of red brick, with a stub of an extension protruding at the back for a tiny kitchen and box room. The front door opened directly onto the pavement. At the rear a steep alleyway ran down towards the distant sea. The back-bedroom window overlooked a verdant patch of allotments below, and higher up stood the Portland Arms pub. Gee's father had been a blacksmith and she had attended a small private school until the age of fifteen. Living all the time at home, she had inspected parts in a local aircraft factory during the war and in 1950, as a clerk, joined UDE. She seemed to have no close friends. Although colourless and unremarkable in appearance, and perhaps at first sight a lifelong spinster, Gee had been engaged to a Scottish carpenter for a period before meeting Houghton. Whyte warned the local police that Ethel Gee must now also be regarded as a suspect.

On his return to London, Whyte found a moving – and what proved to be for their relationship momentous – letter from his girl-friend Bridget. Like many women of her background and generation she had not had the opportunity to study at university, and had few prospects of promotion in a Security Service riddled with sexism. Living on a meagre MI5 salary in a Security Service flat, and spending many weekends with her elderly father, she undoubtedly feared remaining a spinster if David Whyte did not commit to marriage soon. 'Darling David,' she wrote, 'I need you very much and feel that two imperfect halves might make a better whole.' The letter clearly struck home. Soon afterwards, Whyte proposed. Bridget accepted and they started to plan a wedding in early September. They told only a handful of people at Leconfield House.[23]

Whyte needed to manage this personal turmoil while keeping Britain's most important overseas intelligence ally, the Americans, abreast of the Houghton investigation because of the 'Sniper' connection. Whyte had recently visited Harold Shergold's office in MI6's Broadway, London, headquarters to discuss the issue. They had agreed that the latest developments in the Houghton case, in Shergold's words, went 'a long way towards establishing the reliability of ['Sniper'] information', and MI6 pressed for more to be divulged to the CIA on the progress of the 'Reverberate' investigation. The background Shergold confided was that the CIA's ever-suspicious head of counter-intelligence, James Angleton, 'has always maintained that one of the weakest features in ['Sniper'] reporting has been that he has given no leads to any current Polish or RIS [Russian Intelligence Service] agents'. MI6's head of station in Washington DC, John Briance, had already been quizzed about the Houghton investigation by CIA headquarters and informed the Americans that it was 'proceeding satisfactorily and that nothing had emerged which so far in any way invalidated the information provided by ['Sniper']'.

In a letter of 28 July marked 'Top Secret and Personal', Shergold renewed the pressure on the Security Service. 'It is a most fascinating story,' Shergold began, 'and you certainly seem to have hit the jackpot.' He revealed that the head of MI6, Sir Dick White, had instructed Shergold 'to stress the desirability of making some factual statement to CIA at the earliest opportunity'. Shergold asked if MI5 would object to the CIA being fully updated on the investigation and was informed that MI5 thought '"Reverberate" still has a current intelligence role'. Shergold stressed that the last point was particularly important because of Angleton's distrust of 'Sniper' material. Whyte knew he would need to respond with care, steering between the Scylla of MI5 caution and the Charybdis of needlessly irritating MI6 when he knew a priority of its influential head was to be helpful to the Americans.

Details about the missing Gordon Lonsdale trickled in. The Royal Canadian Mounted Police replied to MI5's request for information on 2 August, attaching a passport photograph of Lonsdale, which Whyte immediately passed to the A4 watchers. Skardon called back to say 'he is probably the man in question'. At last the Security Service possessed an image of their new quarry. The RCMP memo summarised the skeletal facts known about Gordon Arnold Lonsdale. He was born at Cobalt in Ontario; his father, Emmanuel 'Jack' Lonsdale, was a labourer and his mother, Elena Bousu, had become a naturalised Canadian in 1931. In 1955, when Gordon Lonsdale applied for a Canadian passport, he had said he intended to sail for England to undertake postgraduate academic studies. He had no security record. Whyte requested further information about Lonsdale and his parents, although cautioning the RCMP not to make 'any enquiry in Canada which could possibly alert Lonsdale'. He stated for the first time in writing a thought that had been swirling around in his head for several weeks and which he had discussed animatedly with fellow officers: that if Lonsdale was involved in espionage, he 'may have some illegal role'. The same day Whyte instructed the A4 watchers to start surveillance on the offices of an estate agent at 19 Wardour Street in London's Soho in case Lonsdale visited. It was run by a man called Michael Houlbrooke Bowers and Lonsdale had given it as a forwarding address for his mail. Whyte warned the MI5 watchers that if Lonsdale was an 'illegal' he would be 'highly trained and alert'; 'nothing should be done to alert him and, rather than run the slightest risk of being blown, observation should be broken off'.[24]

KGB spies came in two guises. The vast majority were 'legals', that is they operated out of the Russian embassy and had diplomatic cover. They were closely monitored by MI5's watchers as far as their limited resources allowed, and the Soviet counter-espionage section of MI5 (D1) under its new head, Arthur Martin, had built up an index of known and likely legal KGB officers. A very small number, however, were the renowned 'illegals', agents who operated under deep cover with a false identity and no diplomatic protection. Although their knowledge was very patchy, Western intelligence agencies understood that, with the onset of the Cold War and the McCarthyite crackdown on communism in the USA, the KGB would undoubtedly attempt to recruit or infiltrate illegals in the West. Their suspicions were amply confirmed when a man called Rudolf Abel was exposed by a KGB defector in 1957 to be an illegal Soviet spy operating in New York. The FBI led that investigation and Whyte knew that they had shared their knowledge of how Abel operated with the Security Service because he had been in charge of the MI5 file on the case. There had been a sensational public trial of Abel in America and he had been sentenced to thirty years' imprisonment. Still languishing in Atlanta Federal Penitentiary in 1960, Abel faced twenty-seven more years of confinement before his release.

Whyte understood that he needed an experienced case officer to focus on Lonsdale, and decided to entrust this work to someone he could rely on completely, his colleague and friend in the same section, Charles Elwell. If occasionally rather blunt, Elwell was thorough and well organised, sharp and tenacious. Born in 1919, he had studied Modern Languages at Oxford before joining the Royal Navy in 1940. He landed agents in occupied Europe from motor gun boats until his luck ran out in 1942 off the coast of Holland. Captured by the Germans, he was sent to Colditz after an escape attempt and remained there until the war ended. He joined the Security Service in 1949, and spent several years in Singapore with his growing family before returning to Leconfield House in 1955 and being posted to counter-espionage. Here his investigative skills had been demonstrated three years later in unmasking an electronics engineer as a Czech secret service agent stealing information about an important RAF guided-missile project.[25] Whyte wanted Elwell to concentrate on Lonsdale while Craggs continued to focus on Houghton and Gee.

This is the first time that Elwell's name appears in the MI5 files for the Portland case but it is inconceivable that Elwell had not discussed the investigation before with Whyte. They were long-standing colleagues together in counter-espionage. Elwell and his wife Ann were personal friends (Ann Glass herself had a distinguished career in MI5 before marriage to Charles). They attended social functions together and dined at each other's houses: after supper Whyte would sit at the piano and either play some of his favourite classical pieces or accompany Elwell while he sang popular operatic arias in his fine but untutored tenor voice.[26]

Until this time, Elwell had been focusing on other Polish and Czech counter-espionage matters, and working his contacts in the frontier world of East European émigrés in London. He was a frequent visitor, for example, to the dimly lit and lushly upholstered basement rooms of the Eve Club in Regent Street. It was presided over by the diminutive blonde figure of Helen O'Brien, originally a Romanian refugee, and her husband, Jimmy, who had become friends of the Elwells. The Eve Club was the haunt not only of celebrities such as Frank Sinatra, but also of businessmen, embassy people, young women and – most importantly for Elwell – Soviet diplomats and defectors. Of middle height, handsome (if sometimes a little vain), charming (except when he chose not to be), an excellent dancer and blessed with a good sense of humour and the ability to enjoy himself, the MI5 officer was a stalwart of the club, sipping his favourite tipple (a dry martini) while keeping a beady eye on proceedings. A typical entry in Elwell's diary, dated a few days before his stint as MI5 duty officer for the weekend of Houghton's planned visit to London, noted that he had 'Spent the evening at the Eve Club. Ate frogs for the first time. Not v. interesting. Excellent chateaubriand.'[27]

As Houghton's anticipated visit to London on the weekend of 6 and 7 August approached, Lonsdale was observed at the estate agency in Soho meeting various business acquaintances of dubious appearance, and he was trailed to his new rented flat at 23 Devonshire Terrace in Bayswater.[28]

On the Saturday afternoon, a team of A4 watchers were waiting for 'Reverberate' under the Victorian ribbed girders and high arched roof of Waterloo station when his steam train hissed to a halt. They noted that Houghton was carrying his usual brown leather briefcase and needed to ask directions from 'local lads' before walking to the area near the Old Vic theatre. The A4 officers 'thought that at this time he looked rather restless and fidgety but did not take any action to establish whether he was being followed'.

At two minutes before 4 p.m. Houghton 'took up his position outside the Old Vic and paced up and down until precisely 16.00', when he was approached by a man the watchers identified as Gordon Lonsdale from the previous meeting and the photograph provided by the Canadians. Lonsdale arrived from a street opposite 'carrying a thick newspaper'. After dawdling in a side street for a couple of minutes the two men crossed into Lower Marsh, a narrow jumble of terraced shops and houses which runs down the south-east side of Waterloo station and where there had been a lively street market since Victorian times. The brouhaha of the market 'afforded excellent cover' to the A4 surveillance team. The two suspects halted in a shop entrance before moving on to a 'cheap café, Steve's Café' in Lower Marsh. Here the watchers took a risk. In Skardon's words, 'the senior officer thought it worthwhile having two officers who had remained in the background to partake of refreshment with the possibility of establishing some of the events which would probably take place in the café between Lonsdale and Houghton'. The two suspects sat opposite each other at a corner table. The MI5 officers slipped into a table 'close to and behind Lonsdale (the cagey one)' a few seconds later.[29] One of the officers had his back to Lonsdale, so close that 'every time he spoke I could feel his movement whilst [Lonsdale] was pushing the chair and bending forward'. With linoleum on the floor and the fug of cigarette smoke, the café had seen better days. Over the murmur of conversation from the Formica-topped tables and the hiss of steam from the chrome vat at the counter, and nursing their own mugs of strong tea, the two MI5 watchers caught snatches of the interchanges between Lonsdale and Houghton. They saw Lonsdale examining a newspaper cutting:

> L to H: 'I wonder if this story is correct.'
> H to L: 'Yes I am sure they went over.'

'The conversation here', Skardon's report noted, 'was about the two US maths experts who had defected.' These were Bernon Mitchell and William Martin, who both worked for the National Security Agency (NSA), America's highly secret code-breaking agency. The day before, 5 August 1960, the Pentagon had announced that the two men had not returned from holiday and said: 'there is a likelihood that they have gone behind the Iron Curtain'. Western newspapers were crammed with speculation that Martin and Mitchell had defected to the USSR.[30]

> L: 'You seem to have plenty in your attaché case.'
> H: 'Yes I have more than my sleeping and shaving kit.'
> L: 'We can arrange those meetings if you would like to put them in your book.'
> H: 'Yes I will, they will take some remembering.'
> L: 'These will be the first Saturday in each month, especially the first Saturday in October and November . . . at Euston station. [The watcher added a question mark after Euston because

Lonsdale's 'voice tailed off slightly'] . . . The driver will sit in a car in the area. I don't know where. I'm 90 per cent sure I will be there. We will use an interpreter . . . You will have to find him.'

They then spoke of the benefit of Houghton leaving his car at Salisbury and coming to Waterloo by train:

L: 'The packet looks fat. Seems like a lot of work for me tonight.'
H: (laughingly replied) 'Plenty . . . That room at the hotel is expensive.'
L: 'That will be taken care of.'

The watchers remarked, 'Whereas Lonsdale spoke in rather hushed tones and always craned forward when trying to impart some information secretly to Houghton, the latter spoke in rather a loud voice and made no pretence of secrecy.' Houghton continued by saying he was meeting 'a South African girl tonight' at his hotel and asked if Lonsdale could help her get 'fixed up' with a job. Just before leaving the café, Houghton was overheard to say, 'I don't want paying yet.'

At 4.30 p.m. they walked to a neighbouring road, which led back towards the Old Vic. They halted here, and Lonsdale opened the door of a red telephone kiosk for Houghton. The watcher concealed himself 'behind a broken-down lorry on a convenient bomb site . . . immediately behind the kiosk'. Houghton entered the booth and removed a package wrapped in brown paper from the external compartment of his briefcase. Lonsdale handed him a newspaper, and Houghton slipped the package inside before handing it back to his companion. The two then sauntered back to Waterloo Road, and 'after a final exchange of words and a backslap from Lonsdale they parted'.

The watchers made calls to Elwell in the operations room on the third floor of Leconfield House to discuss what to do next. Elwell said it made sense to concentrate resources on trying to find Lonsdale. MI5 now knew he was renting a furnished room in Devonshire Terrace, but there was no sign of either the suspect or his car there. Houghton meanwhile met his South African acquaintance, Miss Hope Fripp, at the Cumberland Hotel, where they had a few drinks, Houghton 'obviously going out of his way to be very sociable to a fairly attractive forty-year-old'. MI5's checks on Fripp, however, had not suggested anything suspicious. The meeting appeared to be nothing more than Houghton using one of his London trips for socialising. He left early the next morning to return home.

When the A4 watchers' shift arrived at 8 a.m. on Sunday at Devonshire Terrace there was still no sign of Lonsdale or his car. Elwell walked home from Leconfield House to Porchester Terrace through Hyde Park, and then took various phone calls from A4 asking for advice. Officers scoured London for the elusive Canadian. Checks were made at the car park at London airport at Heathrow (Gatwick had opened in 1958 but was still of little significance in 1960) in case Lonsdale had travelled – even perhaps fled – abroad, and at addresses in London linked with him. There was no sighting. Lonsdale had vanished again.

II

Whyte decided it was time to regroup with Elwell. Even though Lonsdale's whereabouts were unknown, a key priority was to learn as much as possible about him. He was given the code-name 'Last Act'.[31] A tap was placed on the phone of the estate agency in Wardour Street, which was also the registered address of the jukebox company which Lonsdale was involved with, and mail to and from him at the same business address was intercepted. An observation post was set up opposite Lonsdale's home address in Bayswater. Elwell set to work building up a picture of Lonsdale and his business activities. The A4 watchers fleshed out the description of the mysterious Canadian: 'Looks thirty-five. 5ft 8/9 ins. Medium build, inclined to be heavy. Crown of head has Slavian straightness. Dark-brown wavy hair, full round face, obese trend, sallow complexion, dark eyes with slight slant, broad nose nearly pugilistic flatness, walks with hands clasped in front of him. Canadian. Dressed in single-breasted dark-blue suit, white shirt, dark-blue tie, black shoes. Drinks beer. Has quiet Canadian brogue but with slight foreign accent.' Lonsdale was shadowed on a visit to Bury St Edmunds in Suffolk, where he and a clutch of business associates met a company marketing an anti-theft device for cars.[32]

The 6 August meeting between Houghton and Lonsdale convinced MI5 that Houghton was actively spying, and that Lonsdale was an illegal resident (the illegal head of a network of agents) for the KGB or the intelligence service of a Soviet satellite country. Lonsdale might be controlling more spies than Houghton and, as Whyte warned the Admiralty, 'indeed, more important ones', and as a result the Security Service wished to let the investigation run on for several months. Gee might be supplying Houghton with sensitive information and he may even have recruited 'other sub-sources in the Portland area'. On 11 August Whyte sent a 'Top Secret' letter to MI6, briefing them on all the important facts known at that time about Lonsdale. But the key purpose of the letter was in the last paragraph: 'It is possible that we may make some urgent request for action by one of your overseas posts if Lonsdale leaves the UK.'

While Elwell went on a family holiday to the south coast, James Craggs chased down various potential leads on Houghton. As the information trickled back it seemed clear, for example, that the woman called Nancy Elphick was not implicated in Houghton's espionage. The Dorset Special Branch made enquiries about another of Houghton's girlfriends, Kath Small. She had been married at one time to a South African, so Craggs speculated that there might be some connection with the South African woman, Hope Fripp, whom Houghton had taken out on his most recent visit to London.

On the Wednesday of the same week, 17 August, the Deputy Director of Naval Intelligence, Tony Pigot, travelled down from London to Portland to assess how crucial the information held, and the work done, at UDE really was. Pigot summarised the headline points of what he had gleaned in Dorset by phone to David Whyte. By 'far the most sensitive information' held by UDE was the 'Top Secret Submarine Intelligence Reports'. The declassified MI5 documents do not make clear the nature of these documents. But they undoubtedly covered the intelligence-gathering forays British submarines were making in the greatest secrecy into Soviet territorial waters at this time. Sonar tapes collected by these 'mystery boats' were sent to Portland for analysis.[33] The reports, however, were 'very closely held' in one UDE office to which Houghton and Gee could not have access. Pigot also referred to the important '2001 project' to develop a new type of sonar, known then as ASDIC,[34] for the UK's first nuclear-powered submarine, HMS *Dreadnought*. This was not yet in production but was being tested at Portland. The work, Pigot told Whyte, 'so far as UDE are concerned, is not particularly sensitive' because there was 'no really staggeringly new feature in it, except a somewhat increased range'. It was classified 'Secret', and 'Miss Gee handles all the drawings of it'. Pigot was far more anxious about the '2001 project' than UDE: it 'might be of quite considerable interest to the Soviet Bloc to know how far we had progressed [with it].' Houghton had access to various other confidential Admiralty papers, but these were 'of a fairly routine nature'.

The other important points to emerge were that the Admiralty agreed to the 'Reverberate' investigation continuing until after the possible meeting between Houghton and Lonsdale at the start of November, and 'not to make any attempts to tighten up on the security arrangements at UDE or elsewhere'.

Meanwhile, Whyte needed to manage the relationship with the Americans. He updated the deputy CIA station chief in London, John Caswell, on the 'Reverberate' case.[35] The Admiralty had discovered that an American scientist was working at UDE, but fortunately his activities were not 'highly sensitive' and there was no reason to think Ethel Gee had access to information about them. Harold Shergold of MI6 had heard of this development. He telephoned Whyte urgently, because in his letter to the CIA at the start of August he had not only told the Agency what had been agreed about the Portland investigation but gone further than MI5 wished, stating that 'American information was not in any way endangered in this case'. Shergold and Whyte agreed that this statement needed to be qualified by another MI6 letter, adding a caveat about the US research scientist. MI5 and MI6 knew that the link with the Americans was too valuable to be compromised by an over-optimistic and premature desire to provide false comfort to a friend.

When Peggy Johnson's statement arrived by diplomatic bag from Malaya in early August, Brian Wise emphasised that 'she would prove a very poor witness'. Her highly wrought emotional state and fixating 'around and around ad infinitum' on some topics (especially Ethel Gee, whom she described as 'a dominating personality, not bad looking, but extremely fat') meant that the statement took three hours 'to extract'. MI5 would be foolish to rely on Peggy Johnson to be a lead witness in any prosecution and clearly needed much more evidence.

August slid by for Elwell and the team investigating Lonsdale.[36] The mail to and from Lonsdale's flat at 23 Devonshire Terrace was tested for invisible ink (with no success); MI6 passed copies of Lonsdale's photograph to its sister intelligence agencies on the Continent in case he fled abroad, and asked its equivalent in France to investigate a contact of Lonsdale's in Paris; the watchers kept under surveillance Lonsdale's meetings with some of his louche business associates in Soho and south-east London and various girlfriends, including some elegant young Italian women.

On the afternoon of Friday, 26 August, however, Elwell was surprised by a phone call from Jim Skardon of A4. It was propitious because unexpectedly it was this call that was to lead MI5 to the first irrefutable evidence that Lonsdale was an active KGB illegal spy. The watchers had followed Lonsdale in his car from Paddington to a branch of the Midland Bank in Great Portland Street, where he had deposited his briefcase, attaché case and cash box, and told staff he would be leaving shortly for Canada and gone for just over a month.

Within days the MI5 covert operation to search Lonsdale's bank deposit box was agreed in principle for Monday, 12 September. The timing could not have been worse for David Whyte. He was about to take leave to celebrate his wedding to Bridget on Saturday, 10 September. They wished the event to be as private as possible and kept it secret from colleagues at MI5. Whyte organised the marriage at St Mary's Church in Bryanston Square, near his house, and made arrangements for the honeymoon. Only two witnesses were to be present at the ceremony, both from MI5.[37] The marriage passed off smoothly and the newly married couple drove off immediately on their honeymoon, motoring down through France to Italy to revel in the late-summer sunshine of southern Europe, from where Whyte sent postcards to MI5 colleagues to announce his marriage. The wedding, of course, explained why Whyte was forced to leave the highly sensitive operation to search Lonsdale's bank deposit in the hands of Elwell, and why he was absent from the MI5 laboratory when the stunning discovery of Lonsdale's KGB espionage equipment was made.

Quite a retinue of MI5 officers wished to share the excitement and took part in the bank operation, in addition to Elwell: the Director of A Branch (General Services), Colonel Malcolm

Cumming; and the head of A2 (technical services), Hugh Winter-born, and his team, including MI5's first scientific officer, Peter Wright. Winterborn was a former army officer who had seen service in China and Japan (learning to speak those languages fluently) before joining MI5. Working in tandem with Wright, he organised with flair, military precision and humour the Security Service's clandestine operations, in particular those of dubious legality, such as buggings and burglaries. Wright had been recruited in 1955 to advise A Branch on the operational use of electronic and other scientific equipment. He was a tall and striking presence, stalking the corridors of Leconfield House with his shock of white hair and outsider's love of challenging the Security Service's long-held and comfortable conventions.[38] The discovery of the cipher pads, the map of roads in the Kingston area, photographic equipment and a hoard of photos and documents proved beyond reasonable doubt that Lonsdale was a Soviet illegal. MI6 and the CIA were quickly informed, 'because of the bearing that this development had on CIA's assessment of the bona fides' of 'Sniper'.[39]

Elwell and his colleagues worked feverishly to assess the various items found in Lonsdale's deposit box. Among these were the stash of Lonsdale's photographs. On the evening of Thursday, 15 September, Elwell was, he wrote later in his diary, 'examining the belongings of a Canadian named G.A. Lonsdale who was suspected of being a Russian Intelligence Officer and whom I had been directed to investigate'. To his astonishment, he

came across a visiting card of an army officer on which was written on the back the name Elizabeth Farquhar Oliver and our address in London, 5 Porchester Terrace. Elizabeth FO was . . . Lou Oliver [a family friend of the Elwells]. Being rather struck by this discovery I telephoned my superior at the office [Martin Furnival Jones] and told him about the visiting card. He said: 'Are you sure you did not know Lonsdale?' I said I was sure. Soon afterwards one of my assistants brought me a large batch of snapshots and asked whether they should be copied. I looked at some of them [and] thought they were of no significance but directed that they should be copied. 'They might contain something of interest,' I said. Soon afterwards I went home.

Elwell's diary entry continued:

> The next morning as soon as I arrived at the office I was sent
> for by ... Furnival Jones. He said, 'Are you quite sure you did
> not know Lonsdale?' I said I was. 'Then how', he said, 'do you
> account for the fact that a photograph of you was found yesterday
> evening among his things?' I said I could not account for this and
> suggested I should go off at once and have a look at it. He agreed.

Elwell was shown a photograph of himself with

> a rather pretty girl ... I recognised it at once as one of me with
> Lou Oliver and as being taken in the flat at 5 Porchester Terrace. I
> cannot remember whether I immediately recollected the occasion
> as having been Tom Pope's party of students. [Tom Pope was a
> young Canadian diplomat studying at SOAS who rented the flat
> at the rear of 5 Porchester Terrace.] Anyhow I telephoned [Fur-
> nival Jones] and suggested he shd. ring A. [Ann Glass, Elwell's
> wife, then working in the Foreign Office] at her office and ask
> her to come over so that she could corroborate me without being
> able to prompt her. He said rather nastily: 'How do I know you
> have not been in touch with her already?' I said there were plenty
> of people present in the room to prove it.
>
> However he did ring A ... Graham Mitchell sent his car for
> her and she was shown up to him and shown the photograph
> which she at once saw had been taken at the flat ... she explained
> that it must have been taken at Pope's party by a Canadian [Lons-
> dale]. She recalled how she had asked him why he had come to
> study Chinese in the UK when the facilities in N. America were
> so much better and how he had abruptly broken off the conver-
> sation and moved away.

Lonsdale had been invited to Pope's party as a fellow student of
Chinese, and had gone around taking photos of the guests. Among
his snapshots were others of Ann Elwell, and this all helped to corroborate the Elwells' explanation. Elwell considered it was 'unique
in the annals of counter-espionage' that a spycatcher investigating
an illegal should have been photographed unknowingly by his
target.[40]

At around the same time the Security Service learnt that Lonsdale had left other belongings at a depository, Hudsons, near Victoria station on the morning of Saturday, 27 August, before he went abroad. Jim Skardon visited on 14 September. Its director (described in Skardon's note of the meeting as 'a rough diamond and a typical self-made businessman ... [who was] no stranger to hard liquor') allowed the Security Service to remove the items for examination. In great secrecy MI5 transported Lonsdale's possessions from the depository to the MI5 laboratory, where Elwell examined them with his technician colleagues that afternoon and the next morning.[41] The most significant discovery was a Chinese wall scroll: X-rays showed that the lower roller was not solid but hollow, and after various experiments probing with a pin based on their knowledge of similar Eastern Bloc devices, it sprung open and could be unscrewed. There was nothing inside. It was, he thought, a 'cunningly contrived hiding place such as is used by the Russian Intelligence Service ... [which] would be ideal for hiding documents received by Lonsdale from his agent'. Other finds of potential importance were a Bush radio, suitable for receiving short-wave messages from Moscow, and undeveloped film from a miniature Minox camera. Unfortunately, when MI5's locksmith, Jagger, was relocking one of the items, a Samsonite suitcase (ironically containing nothing of interest), his tool slipped near the lock and gouged a deep scratch about half an inch long. The technicians tried to age the scratch but it remained the only one anywhere near the lock.

Having caught so much new information in the counter-espionage net, a crucial task was to sift and analyse it. Although MI5 employed a number of remarkable women doing officers' jobs in 1960, a meagre handful were of officer status. One was a researcher in the Soviet counter-espionage section, D1, Evelyn McBarnet. She was a long-standing friend of Ann Elwell from her days in the Security Service and had worked with Ann on furrowing through the Burgess and Maclean files after their defection in 1951. McBarnet had devoted her life to the Security Service, living and breathing its history and gossip. She was renowned for her fierce intelligence, keen eye for detail and organisational skills, had a notable birthmark running down one side of her face and worked directly as research officer to the Director of D1, Arthur Martin.[42] McBarnet set to work, creating files on the varied aspects of Lonsdale's life: from his mistress in Brussels, Denise Peypers (discovered from Lonsdale's intercepted correspondence), and various girlfriends – including an Italian and a Yugoslav – to the companies he was involved with, and his travel and accommodation. She also prepared notes on Lonsdale's contacts and a detailed biography of Lonsdale himself. On 16 September Graham Mitchell visited 'C', Sir Dick White, in his Broadway office to brief him on the Lonsdale case, because of its bearing on how reliable intelligence from 'Sniper' was regarded and on relations with the CIA – 'a matter of very considerable importance to MI6'. White 'evinced the liveliest interest' in Mitchell's narrative.

The list of eight roads in south-west London found in Lonsdale's cigarette lighter was regarded as highly significant: almost certainly locations where Lonsdale met his espionage contacts. Extraordinary plans were put in place to create an average of two observation posts seven days a week in each of the eight roads, to be manned from 7 a.m. to 10 p.m. in readiness for Lonsdale's planned return on 26 September. They were to be staffed by a special group of about 100 people, recruited from Leconfield House, former members of the Security Service and police officers. Women were to be preferred. Malcolm Cumming, Director of General Services, and renowned for playing the squire at his extensive country estate in Sussex and the spy in town, noted that 'experience has shown that householders generally preferred women to men in OPs, particularly during daytime when the lady of the house was usually alone in charge. Furthermore, the presence of a woman (e.g. 'niece from the country') tends to attract less curiosity from neighbours.'[43]

Lonsdale's spy equipment at the bank and his impending return to Britain sparked a tense debate at the highest levels of Leconfield House. Should the Security Service covertly examine 'Last Act's espionage paraphernalia for a second time? Should MI5 arrest him immediately after he had collected his possessions from the bank? Should it risk imposing surveillance on Lonsdale on his return?

Elwell and his colleagues were acutely aware of the risk of Lonsdale spotting that his possessions in the bank had been tampered with. Not only was there the broken rubber band, but when opening the Skyline attaché case which contained Lonsdale's lighter the lock had been slightly knocked out of shape, and there was a risk that Lonsdale could have set a trap by leaving unexposed 35mm film protruding from the Praktica camera. These risks could be reduced by a further examination (for example, by substituting a perished but unbroken rubber band), and there was a huge potential dividend in new intelligence to be gained from developing the various camera films, and especially from photographing more of the cipher pads. Furnival Jones met with Elwell, Leggett and others to discuss the issue on 21 September. One cipher pad was in current use, Leggett wrote after the meeting, which 'is probably "Last Act's" incoming traffic pad on which he receives coded messages from Moscow; if we can identify the particular messages – as we have a very good chance of doing – then we should be able to break and read as much of "Last Act's" incoming traffic as we have pad for. At the moment we have photographed 680 groups of this pad which might give us about five average messages. If we can photograph the remainder of the pad, we estimate this would give us about 4,400 more groups.'

From the number of pages already used, MI5 and GCHQ, Britain's code-breaking agency, estimated that 'Last Act' had probably been at work for 'at least two years'. The Security Service files confirm that MI5 had access to a cipher pad very similar to the ones Lonsdale held in the bank. Peter Wright knew from attending meetings of a group with the deliberately dull title of the Radiations Operations Committee (which in fact led crucial work coordinating the increasing number of British signals intelligence interception operations) that the Swiss intelligence service had recently discovered a cipher pad abandoned by the KGB. MI6 asked the Swiss if they would urgently lend it to them. They agreed and it was flown to Heathrow by an RAF plane specially chosen for the mission. Wright drove out to meet the plane and was relieved to find that the Swiss cipher pad was remarkably similar to Lonsdale's, held together by a gossamer-thin film of gum. With scrupulous care, the sheets of the Swiss pad were prised apart in the laboratory, photographed, and the glue analysed. It was not made in the West, but Post Office special technicians told MI5 they could create glue almost indistinguishable from the original. As a result the meeting on 21 September was informed that 'the operation of dismembering "Last Act's" OTPs, photographing them, and then sticking the edges together again with gum would carry little risk of arousing "Last Act's" suspicions'.[44]

These decisions were so weighty that MI5's Director General, Roger Hollis, held a summit meeting the next day with his deputy and Furnival Jones. Hollis decided a second operation should be mounted on Lonsdale's attaché case at the bank, but very sensibly ordered that it should be done under the cover of a Special Branch search warrant, so the evidence collected by the police could be used in court and the legal position of both the bank and MI5 would be protected. Hollis accepted Furnival Jones's recommendations that Lonsdale should not be arrested immediately unless the second bank operation 'went seriously wrong', and that Lonsdale should be put under surveillance on his return. Hollis himself visited the chairman of the Midland Bank, who gave approval for a second operation. After an 'entirely off-the-record' discussion with MI5, the Director of Public Prosecutions, Sir Theobald Mathew, agreed to the search warrant. The risky operation was to take place over the weekend of 24 and 25 September, when the bank would be closed. Lonsdale was due to return the next day. There was no room or time for error.

III

At noon on the Saturday, Superintendent George G. Smith of Metropolitan Police Special Branch and the branch manager of the bank arrived at the MI5 laboratory with the attaché case. In September 1960 the Security Service still did not officially exist. MI5 remained in the shadows, and when it needed a public face to prosecute spies it was forced to rely on Special Branch. Bespectacled and burly, Smith was highly experienced in espionage matters and known as 'Moonraker' Smith because of his West Country background. His avuncular manner concealed a considerable ego, boosted by his courting of journalists to burnish his reputation as a spycatcher – to the chagrin of MI5 officers who carried out the often tedious counter-espionage investigations for which Smith took the credit when completed. FBI headquarters had flown over to London a special set of skeleton keys, and there was a collective sigh of relief as one of them opened the locks of the case easily. When developed, some film from a miniature Minox camera revealed only pictures of a girlfriend of Lonsdale.

Copying started of the red section of the OTP in use. It went 'very slowly', according to Hugh Winterborn's memo on the operation. The cipher pad needed to be prised apart with great delicacy and each of the red and black pages photographed. A start was made on the larger, unopened pad, but the laborious pace of the work meant that only a further six red sheets could be copied. Late that evening a specially created jig squeezed the original pages together so the edges could be recoated with the glue concocted by the Post Office technicians. There was some tension overnight over whether the gum would dry satisfactorily, but when examined the next morning it was pronounced a success. The perished rubber band was replaced, the case repacked carefully, locked again and returned to the bank. The OTPs demonstrated beyond doubt that Lonsdale was in radio communication with Moscow. MI5 knew that if it could intercept signals being sent to Lonsdale from Moscow, armed with copies of Lonsdale's pads, GCHQ would probably be able to decrypt them. Unfortunately, there was no radio schedule detailing when and at what frequency Lonsdale would listen out for messages transmitted to him.

With Lonsdale's anticipated return, the twitchiness within the Security Service was so intense that the watchers were banned from communicating by radio within certain specified areas in London. Monday, 26 September, arrived. Surveillance began. The week passed. There was no sign as yet of the mysterious Canadian.

Charles Elwell – having recovered from the shock of discovering the photograph of himself in Lonsdale's belongings – and Evelyn McBarnet continued their painstaking work throughout September and early October in building up a picture of Gordon Lonsdale.[45] In Canada the RCMP ferreted through official documents and despatched officers to interview people who had come into contact with Lonsdale. The birth of Gordon Arnold Lonsdale in 1924 was fully documented, but the Canadians could not confirm whether he was the same person being investigated by the Security Service in London. Vancouver was where the known facts about Lonsdale began, when on 29 November 1954 he was issued with a Canadian driving licence at an address in the city, 1527 Burnaby Street. This accommodation was demolished soon afterwards and Lonsdale moved to a boarding house. Lonsdale was recalled by the owners, when interviewed by the Mounties, as reserved, unmarried, quiet and paying his rent promptly.

On 21 January 1955 Lonsdale was issued with a Canadian passport. He gave as his permanent address a house in Toronto. This, the RCMP pointed out, was where a couple lived named Bacchus, who were communists. Mrs Bacchus was linked to another member of the Canadian Communist Party who had introduced to the KGB a man who was later recruited as a spy. Elwell commented that these connections suggested that 'Lonsdale's lodgings in Toronto were deliberately chosen'. On 22 February from Toronto Lonsdale joined the Royal Overseas League – clearly hoping this organisation, whose patron was Queen Elizabeth II, would assist him in making contacts in Britain – and the same day at Niagara Falls crossed over into the United States and travelled to New York. On 3 March 1955 he boarded the SS *America* and steamed across the Atlantic to Southampton.

It was here that MI5 and the British police were passed the investigatory baton. On his arrival in London Lonsdale first stayed at the Royal Overseas League in fashionable St James's Street, before renting a small flat in an anonymous residential block called the White House in Albany Street near Regent's Park, where he stayed until August 1958. The Royal Overseas League, Elwell noted wryly, did much to launch Lonsdale 'on his stay in the Mother Country by providing tickets for the House of Commons, the Garter Ceremony at Windsor Castle', and acting as a referee when renting his first flat. Lonsdale was abroad from mid-June to mid-August 1955, and for part of this time he travelled on a bus tour of Europe, meeting various people with whom he stayed in contact and whom MI5 subsequently traced. From October 1955, back in London, he followed two courses in Mandarin Chinese at the School of Oriental and African Studies. One of Lonsdale's fellow students here was Tom Pope, who took the photo that had caused Elwell so much discomfort on 15 September 1960. At Elwell's request, the FBI checked the records of two schools in California where Lonsdale said in his SOAS application that he had studied. Neither school had any trace of a student of that name.

MI5 traced various trips Lonsdale made in Britain and abroad in 1957. One was a British Council vacation course in Torquay, which he attended in the company of an Italian girlfriend, Carla Panizzi. Lonsdale's love of female company and his louche lifestyle were clear. By contrast, the source of his income was a conundrum.

Lonsdale had an account with a Swiss bank, which sent funds of unknown origin to a bank in Canada where Lonsdale had another account.

Elwell learnt that, after rejecting the idea of working as a travel agent, Lonsdale had entered the jukebox and vending machine business. The popularity of coin-operated jukeboxes was at its zenith in the late 1950s. In 1956, sensing a business opportunity, Lonsdale had bought two Minstrel jukeboxes from a company in Kent for £500 and sold them at a profit. He purchased two more a year later and sited them at addresses in south London.

While working with this company Lonsdale made friends with Michael Bowers, who represented the firm of estate agents in Soho which Lonsdale used as his forwarding address. Lonsdale, Bowers and others agreed to fund the manufacture of a slot machine whose unique selling feature was simultaneously to dispense picture cards and bubble gum. The business went bankrupt in October 1959. Lonsdale bounced back and in 1960 joined two firms marketing newly invented electronic locks for cars. One of the firms was based at 19 Wardour Street, and Elwell investigated its directors and Lonsdale's associates with his usual diligence and attention to detail. They were a motley crew, some with slightly tarnished reputations after brushes with the law, including Bowers, who had been suspected by the Dutch police of unlawful business dealings; a Scottish woman, Molly Baker, who had been charged with obtaining credit by fraud and was to play a small but important role later in MI5's investigation; and an Irish garage owner, through whom, just before Lonsdale left England on 27 August, he had acquired a swish blue-grey American Studebaker in exchange for his old car.

Despite their best investigatory efforts, apart from Lonsdale's birth certificate, the Canadians had been unable to obtain any information about him from the alleged date of his birth in 1924 until his appearance in Vancouver thirty years later. His early life was, as Elwell summarised in a top-secret biography of Lonsdale in October 1960, 'shrouded in darkness . . . This total absence of documentation is perhaps the most revealing piece of evidence that Lonsdale is an illegal intelligence agent.' To reach this conclusion Elwell also relied on other points: Lonsdale applied for his Canadian passport on the strength of a birth certificate alone, without any guarantor ('the classical method of the "illegal" to obtain a passport, at all events in Canada'); 'if [Lonsdale] is in fact a Canadian he has severed or obliterated all traces of contact with anyone who may have known him in Canada'; and 'the sources of [Lonsdale's] income are mysterious and his standard of living is higher than could be accounted for by the very modest amount of money' from his business.

Another key task for Elwell was to learn all he could about previous Soviet illegals who had been planted in the West. Two of the most important cases centred on Canada and the United States. In November 1951, thirty-one-year-old Yevgeni Vladimirovich Brik had landed in Halifax, Nova Scotia, with instructions from Moscow Centre to take up residence in Montreal before moving to the United States. Brik began a tempestuous affair with the wife of a Canadian soldier. She persuaded him to make a confession to the RCMP, who in turn decided to run him as an agent (code-named 'Gideon'). 'Gideon' proved to be a fruitful source of information for Canadian intelligence, and the RCMP acquired details of how a Soviet illegal operated. Brik was a particularly taxing agent to handle, especially after his lover ended their affair and he sought solace in alcoholic binges. 'Gideon's career as a Canadian agent ended suddenly in August 1955 when he returned to Moscow for a holiday and disappeared. He had in fact been betrayed by a KGB agent in the RCMP.[46]

The second case, which had attracted headlines around the world, was that of Rudolf Abel, who, as mentioned earlier, was arrested by the FBI in New York in June 1957. As a result of the publicity, Elwell and every other D Branch counter-espionage officer would already have been familiar with the published facts of the case. But MI5 had compiled its own files on the FBI investigation and on how the Security Service followed up the implications of Soviet illegals operating against Western countries.[47]

When Abel was arrested the FBI discovered various items connected with his espionage work as an illegal, including three short-wave radios, cipher pads, cameras and film for producing microdots. In Abel's hotel room the FBI found 4,000 dollars, an extended antenna hung out of the bathroom window to receive messages from Moscow on his short-wave radio, a Russian cipher book and valuable material on microfilm (including his 1957 radio receiving schedule, and letters from his wife and daughter). There were intriguing parallels with Lonsdale's spy equipment in the bank and depository. Also uncovered were photographs of two other suspected KGB agents and recognition phrases to establish contact between agents who had never met before (known as 'paroles').

The Abel case highlighted how Western intelligence agencies faced great difficulty in detecting illegal activity, and that it was essential that they should therefore cooperate and exchange information. This arrest revealed new information about KGB undercover espionage tradecraft, in particular the means of communication Abel used, such as encoded radio messages, microdots and also what were known as 'flash transmissions': a method of sending encoded messages by radio in a rapid burst, so the chances of interception were correspondingly reduced.

Before the autumn of 1960, when they started investigating Lonsdale, MI5 had no practical experience of KGB illegals in the UK. But the MI5 files on the Abel case demonstrate that the Service had been actively fascinated by, and accumulated a bank of information about, illegals before this date and that the Security Service had not been complacent about their activities and the threat they posed.[48] A central feature of those activities was covert radio communication with Moscow Centre. Fortunately, GCHQ had experts in the area.

IV

When MI5's suspicions of Lonsdale being a deep-cover Soviet spy deepened, it immediately alerted GCHQ. The staff at GCHQ had completed their move from rather ramshackle wartime buildings in various locations to their pristine new offices at Cheltenham just a few years before, and were imbued with a strong sense of pride and of being special, separate from the main Civil Service. Contrary to the popular image of the Cold War successor to Britain's code-breaking centre at Bletchley Park, GCHQ was not a graduate organisation. Many of its (largely male) staff were born locally and had stopped their formal schooling at the age of eighteen, but had become experts in their area.[49] Whyte and Elwell's senior contact at GCHQ was Arthur 'Bill' Bonsall.[50] After graduating in Modern Languages from Cambridge, Bonsall was recruited in 1939 to work at Bletchley Park. By 1942 he effectively found himself in charge of German air force analysis. One of his many initiatives was to fly

RAF aircraft over Germany equipped with radio receivers to map German air defences (a method later taken up by the USAF during the Cold War to assess Soviet and Warsaw Pact air defences). Bonsall had joined GCHQ after the war and rose steadily up the hierarchy. In 1955 he was made the first head of the newly created J Division, responsible for signals intelligence on the USSR and Warsaw Pact countries, and from April 1958 of Z Division, in charge of policy on all intelligence and reporting. Bonsall was at heart a shy man, with a reserved and austere working style and no small talk, but he was thoughtful, confident in his judgements and preferred to work by personal contact rather than on paper. He was blessed with a remarkably sharp intellect: taking a problem to him was like 'an Oxford tutorial with a kindly don', one colleague later confided.[51]

Bonsall undoubtedly met and discussed the case with the Director of GCHQ's J Division, Teddy Poulden, because in 1960 this department housed GCHQ's special section devoted to Soviet agent radio broadcasts (including the 'flash' transmissions highlighted by the Abel and Brik cases), known as the HF section because these broadcasts were high-frequency.[52] During September MI5 debated with GCHQ how the agencies might trace and decrypt Lonsdale's radio signals on his return to the UK. At the end of September, after the second search of Lonsdale's attaché case, the Security Service in great secrecy sent the photographs they had made of the KGB cipher pads found in Lonsdale's deposit box to J Division and the HF section. They may well have been despatched by GCHQ's special courier service, which ran daily between London and Cheltenham: two vans loaded with highly sensitive material left GCHQ's London office in Palmer Street near St James's Park and Cheltenham simultaneously at 6 p.m. every evening and met halfway at a secret rendezvous point. Here the two drivers swapped vehicles and drove their confidential cargo on to its final destination.[53] Once the copies of the pads were safely in Cheltenham, GCHQ's cryptographers started work.

The painstaking assessment of Lonsdale's belongings found in the deposit box continued: document was cross-referenced against document, paper against paper. At the end of September Evelyn McBarnet spotted a hotel bill receipt for the night of 28–29 June 1960 at the Crown Hotel at Blandford Forum in Dorset, near Portland. She recalled that Houghton's neighbour, Boggust, had observed that

'Reverberate' had a visitor that evening who resembled Lonsdale. Boggust could not positively identify Lonsdale from the photograph but the Security Service were convinced that the mysterious visitor had in fact been the Soviet illegal.

MI5 expected Houghton to meet Lonsdale on Saturday, 1 October – a rendezvous foreshadowed by the overheard conversation in Steve's Café in August. The Admiralty was increasingly jumpy about the case running on with Houghton remaining at liberty to disclose more secrets. With possible arrests imminent that coming weekend, the Director of Naval Intelligence, Nigel Denning, wished to alert the professional head of the Royal Navy, the First Sea Lord, and a handful of others. MI5 persuaded Denning to wait. The Director General, Roger Hollis, was that week to decide 'the future conduct of the case'. Denning said he would 'have a word' with Hollis at the regular Thursday meeting of the UK's supreme intelligence coordination group, the Joint Intelligence Committee (JIC). With confirmation of legal advice that the Security Service now possessed sufficient evidence to make arrests, the debate inside MI5 about next steps came to a head. Should the Service be prudent and arrest Houghton (and Lonsdale, if he reappeared) that weekend? Or should it let the case run on? Charles Elwell in particular, leading the investigation into Lonsdale, emboldened by the discoveries in the deposit box of the illegal, argued strenuously against what he regarded as premature arrests.[54] At the JIC meeting Denning took Hollis to one side and asked him bluntly whether the Security Service were planning to make any arrests that Saturday. Hollis said no, emphasising that the case was 'a very important one' and MI5 were 'extremely anxious to let it run for a time . . . to learn as much as we could about its ramifications'. The DNI agreed to let the investigation continue until November. Hollis stressed to Denning that he should 'keep very much to himself the information about these cases'. It would be 'disastrous', Hollis murmured, 'if any whisper should reach "Reverberate" or "Last Act"'.

Telephone intercepts in Dorset had confirmed Houghton was planning to take leave early in October to visit his seventy-six-year-old father in Lincoln. It seemed likely that he, before visiting Lincoln, would travel to London to meet Lonsdale. With confirmation on 28 September that Houghton had booked a room at the Cumberland Hotel for the night of 1 October, a trip to London became certain.

The surveillance operation began smoothly enough. Houghton was followed from his cottage by one A4 car to the main Salisbury road. It sent a message to another car at a junction further ahead near Dorchester to watch out for him. When the watchers' car reached the junction, however, they learnt that Houghton had not passed by, presumably having turned off at some point. The A4 cars waited patiently in Dorset. Houghton had disappeared. In London observation on the Midland Bank in Great Portland Street and other addresses linked with Lonsdale yielded no sign of him. From around 2 p.m. watchers were stationed near the Old Vic, but again Lonsdale was not seen. A car was despatched in vain to the main road into London in case Houghton was to meet a contact at the pub which his ex-wife had mentioned. Once it was clear that there was to be no meeting at the Old Vic, an increasingly desperate MI5 operations room flooded the Euston station area with watchers. They saw nothing. Around the same time that Leconfield House agreed to stand down the surveillance teams at Euston, at 7.45 p.m. a radio message came through that Houghton had been observed ambling through the main door into the Cumberland Hotel. According to the senior A4 officer there, he 'looked very dejected'. His dejection, real or imagined, was nothing compared with the disappointment and annoyance experienced by MI5. Despite the most intense surveillance, Houghton had gone missing for the better part of nine hours. Later that evening he came down 'dressed in a much smarter town suit' from his hotel room and drank and dined with the South African woman, Hope Fripp, he had met on his previous visit to London.

Tension mounted at Leconfield House in the weeks that followed. Lonsdale had said in Steve's Café on 6 August that he was '90 per cent sure' he would meet Houghton on 1 October, not that his appearance was guaranteed. As the days passed, inevitably Whyte and Elwell speculated whether Lonsdale had become suspicious for some reason and flitted, perhaps alerted by some clumsy surveillance or even by some leak from within MI5. The waiting was made more bearable for Elwell only by his fascinating counter-espionage investigation, and the fact that Lonsdale had said he might be absent abroad for as long as six weeks. Hollis in a private meeting on 12 October professed to believe that the fact that Lonsdale had not returned to the UK 'did not yet greatly disturb us', but as the days passed even his calm undoubtedly became more ruffled.

Once again surveillance of Houghton and Gee became hum-drum.[55] Reports on Houghton from his neighbour, Cyril Boggust, painted a picture of the suspect living in voluntary solitude (many entries read 'came home alone', 'went out alone', 'did not hear him return', interspersed occasionally by 'came home with Bunty', 'went out together', 'returned alone'). The routine of James Craggs, the MI5 case officer overseeing their surveillance, was enlivened, but only marginally, by a trip to the Portland base on 12 October. In Dorchester police headquarters that afternoon, Craggs broached the subject of the failed 1 October surveillance of Houghton. The police hesitantly criticised the watchers for not setting up a control room at police headquarters to exploit their more powerful radio. In riposte, Craggs could not resist mentioning MI5's irritation at a security clanger dropped by the Dorset police: they had booked the MI5 watchers into their hotel in Dorchester on the night of 30 September, the day before the failed surveillance against Houghton, as New Scotland Yard officers. The police agreed this was a 'major blunder'. In a small town like Dorchester rumours could easily have started and reached the ears of Houghton on one of his pub jaunts in the local area.

The intercepts of Houghton's phone calls revealed intimate details of his far from conventional lifestyle and interests – especially for a man of fifty-five in the still staid Britain of 1960 – and his true attitude to Bunty Gee. Houghton was fascinated, for example, by an anonymous letter he received inviting him to a 'disgusting . . . Feast of Bacchus party' in a flat at Weymouth. Houghton revelled in telling his friend Kath Small, who lived in Weymouth, what the MI5 transcriber coyly described as 'all the sordid details' of the planned event, which involved three men arriving with their wives or girl-friends and having sex with multiple partners. The much-vaunted sexual revolution of the 1960s clearly started early in parts of sleepy seaside Weymouth. Houghton, the intercepts showed, went out with Small on a few evenings when he was not meeting Gee and invited her back to his cottage. He discussed Gee with Small over the telephone, telling her in response to a question about when he might marry Gee that 'the evil day had been put off' and that he did not consider going to the dubious party with her because 'you'd be surprised how moral she is as regards anything like that'. Small commented a little spitefully, but perhaps truthfully, on Ethel's appearance: 'I saw her the other day, she's ageing . . . within this last – what – six months she's changed considerably . . . One day you're going to turn round and find she's left you.' Houghton replied: 'Yes, well there is that risk – or is it a risk?'

A couple of days later, at 8.40 p.m., Cyril Boggust heard a car draw up outside Houghton's cottage. He switched out the light and, edging aside the curtain, heard laughter and saw Houghton escort a blonde woman aged about thirty-five into the house. Glued to his window, he observed them at one point in an up-stairs bedroom, where Houghton was showing the woman what appeared to be photographs. He heard her cry out, 'You are a dirty old man!' Boggust watched her being driven away at 12.20 a.m. the following morning, Houghton returning home at about 1 a.m. As Boggust pointed out to the Dorset police, this was the first occasion he had seen 'an unaccompanied woman apart from Bunty at the house' since Houghton's divorce. The attractive woman was later identified as Houghton's girlfriend, Kath Small. She rejected Houghton's sexual advances and their relationship petered out by the end of October.[56]

Telephone intercepts recorded Gee's apprehension at someone being searched unexpectedly at the gate of the naval base, and a lack of enthusiasm about a possible promotion, her apathy being encouraged by Houghton, who repeatedly stressed: 'I mean you're not short of money are you?' The relationship between Houghton and Gee was clearly a singular one. Self-contained, marooned in a humdrum life and sharing a bedroom with her elderly mother, Bunty Gee must have been flattered by Houghton's attentions into starting what appears a lukewarm affair. Lacking privacy at home, and no doubt wary of gossip generated by nosy neighbours in a confined community like Portland, Gee never passed the night at Houghton's cottage but was sufficiently willing to flout conven-tion occasionally to stay with him overnight in a hotel, pretending to be his wife. This arrangement probably suited Houghton well. He could exploit Gee for companionship and as a helpmate for his espionage, and meanwhile continue his somewhat grubby flirtations with other women – of which Gee was certainly not aware.

The monotony of the Portland surveillance was broken on Tues-day, 18 October.[57] An intercept of the office telephone of the estate agent in Soho's Wardour Street where Lonsdale sometimes worked confirmed that he had returned that day and told work colleagues that 'having come off the plane, he walked straight in here'. The long-awaited news brought Whyte and Elwell a sense of almost palpable relief. Lonsdale had been absent for seven weeks – one week longer than the Security Service thought likely. The lurking doubt and apprehension dissipated. The hunt was on again in full cry. The watchers put the Wardour Street building under surveillance from the Falcon public house opposite, their observation post being given the apt code-name 'Gin Trap'. The landlord was a former Guards officer, and he offered the watchers the use of the bedroom of his eleven-year-old son in the attic, whose two porthole-shaped windows looked down onto Michael Bowers's office. The network of OPs at the eight roads in south-west London was reactivated, but it proved an expensive and fruitless exercise: there was no sign of Lonsdale.

Intercepted phone calls between Lonsdale and his work colleagues a few days after his return revealed that he had made arrangements to rent a tiny one-bedroom flat from 1 November. This was on the sixth floor, number 634, of the White House in Albany Street, where he had previously lived in 1958.[58] Lonsdale had, in Security Service jargon, been 'housed'. The management of the block proved very cooperative. Hugh Winterborn's A2 technical team sprang into action. The flat next to Lonsdale's, 633, was commandeered. An officer with the code-name 'Alf' occupied it from 1 November, responsible for what an MI5 memo bashfully describes as 'another operation directed against "Last Act"'. Its exact scope is not clear. At the very minimum it involved drilling a hole through the wall into Lonsdale's flat and installing a listening device.[59] To ensure 'Alf' could carry out his work undisturbed, Lonsdale's movements were carefully monitored and within a few days a special telephone was installed in flat 633 which was directly connected to the watchers' control room. The listening bug started to produce intelligence almost immediately (marked 'Top Secret' and code-named 'Robe'), giving intriguing insights into how the deep-cover KGB agent lived. On 13 November, for example, he invited his Yugoslav girlfriend round in the evening. The transcript painted him as courteous and an entertaining conversationalist, comparing Russian and American systems of education, and talking about the changes he had remarked in Britain since his arrival: 'Six years ago there were men in filthy macintoshes, everyone had shabby clothes, shop windows were shabby and old-fashioned and only one or two shops were centrally heated; it was pathetic.' He criticised the food – the helpings were so small that he had to search on the plate for a steak; sugar rationing had only just ended. There had been changes and improvements in the previous five years.

The top-secret operations at the White House also involved GCHQ because MI5 wished to intercept and decrypt any incoming radio messages from Moscow. This project was dubbed 'Operation Jeremy'.[60] The details are obscure because none of the papers documenting GCHQ's involvement has been declassified. It seems that a GCHQ technician, Arthur Spencer (who may have been agent 'Alf'), moved into flat 633 to begin operations. His presence there was disguised by a woman MI5 officer who pretended to be a new tenant and left every morning to go to work, appearing to leave the flat empty. In fact, Spencer lived there almost continuously, tiptoeing around the floor and scarcely putting a foot outside. Inside the flat, according to Peter Wright, he installed the secret 'Rafter' equipment developed by MI5 and GCHQ in the late 1950s. This worked on the principle that every radio set at that time contained a local oscillator which moulded an incoming signal into a fixed frequency. The oscillator in turn always emitted sound waves as it operated, and it was these radiations which disclosed the presence of a receiver, and the frequency it was tuned to. Spencer used the 'Rafter' equipment to radiate through the walls into Lonsdale's flat, where the Security Service knew the suspected illegal had a radio capable of receiving short-wave broadcasts from Moscow. MI5 placed a tap on the mains electricity supply to Lonsdale's radio connected to a silent buzzer. This was worn as an earpiece by Spencer and warned him if Lonsdale listened in to the Russian capital. When the buzzer sounded, Spencer tuned in the 'Rafter' equipment to discover which frequency Lonsdale was listening to and alerted GCHQ's London base in St James's, which then relayed the signal to GCHQ in Cheltenham by means of an enciphered telex link.[61]

The man charged with handling the messages at Cheltenham was Bill Collins, a higher executive officer in the high-frequency section of Department J, which specialised in Soviet signals intelligence. He was not a graduate or professional cryptanalyst and, indeed, was far from the popular image of cryptanalysts in general. He had served with the army in the Far East before joining GCHQ, and was aged about forty in 1960. Tall and thin, he was a keen rugby player and was calm and easy to deal with.[62]

The first of the messages came through to Collins in early November. Armed with the copy of Lonsdale's OTP made by MI5, the decoders began work but were immediately perplexed and concerned: they could not decipher it. This was a major blow.

The principles of the OTP were first established in the late nineteenth century, and when followed correctly provide a truly unbreakable cipher. It is a technique in which normally each character of the 'plaintext', the message (whether letters or numbers), is first converted into a number. Each of these numbers is then encoded using a randomly generated set of numbers before the message is sent. The signal is decrypted by the recipient using a matching OTP. The name OTP refers to a tiny notepad or microfiche containing sheets of random digits or letters, usually (and as with Lonsdale's pads) printed in groups of five, and used to ensure totally secure communications. They had been widely employed by the Soviets during the Second World War and afterwards. To enable a message to be deciphered successfully, the sender must include what is known as an indicator group: a block of numbers on the relevant page of the OTP, often not itself enciphered, which the recipient uses to position the message at the correct place on the pad to decrypt it. With Lonsdale's first message, the GCHQ cryptographers could not locate any indicator group.[63]

One possibility that occurred to the Security Service was that Lonsdale feared his ciphers were compromised and had returned from abroad with a new set of pads. A burglary of his flat would enable MI5 to establish which OTP Lonsdale was using, see if anything new was hidden in the cigarette lighter or Chinese scroll, and gather any other useful intelligence from his belongings. On 10 November Lonsdale was confirmed to have travelled up to Bury St Edmunds in Suffolk on business. At 10 a.m. the manager of the White House opened the door of Lonsdale's flat with a pass key and he, Elwell and Winterborn crept inside. It was small and depressingly bare, the windows wreathed by net curtains, with scarcely room for the bed – a flat for a single man who was a bird of passage. They found the cigarette lighter in a cupboard, took the Chinese scroll down from the wall above the bed and swiftly smuggled them into the flat next door to be examined and photographed.

The scroll was empty. There were three changes to the contents of the lighter since September. The list of eight roads was missing. With a sigh of relief they found the same cipher pad still in use, but now with a red page torn away and a different number group on the top left-hand corner. This indicated that 100 number groups had been used since MI5 had secretly examined the pad in September, and that Lonsdale had started to decipher the most recent incoming message at a different place on the OTP than expected. Also, underneath the red OTP page was an unexpected treasure: a tiny photostat about two by one and a half centimetres, which proved to be Lonsdale's signal plan giving the dates, times and wavelengths of Moscow's wireless transmissions. All the items were replaced in the lighter, and the baize pad quickly re-glued with the help of a hairdryer. Also found in Lonsdale's flat was a new Grundig short-wave wireless set and a photographic enlarger. Elwell rifled through Lonsdale's papers and chanced on a 'short but exceedingly sweet' love letter from Lonsdale to his mistress in Brussels (with the unromantic prudence of an experienced secret agent, he had ordered her to return all his letters). 'Darling,' he had written on 3 November, 'My trip was very successful financially and therefore as soon as I can straighten things out here I will come to see you. I missed you terribly.' The search was over in forty minutes. The new information about the OTP was passed to Bill Collins in Cheltenham, and there was huge relief when GCHQ successfully deciphered the first message. From then on GCHQ were able to intercept and read some of Lonsdale's bi-weekly traffic from Moscow.[64]

After his reappearance in London on 18 October, Lonsdale did not stay at the Albany Street flat immediately. Every evening he walked out of his office in Wardour Street and disappeared. Nor, even after 1 November, did Lonsdale spend every night in his White House flat. The watchers followed him but were hobbled in their work by new restrictions on their surveillance methods. MI5's specialist in eavesdropping technology and later leading conspiracy theorist, Peter Wright, had collected evidence suggesting that the Soviets were able to monitor the radio communications of the watchers, and (much less conclusively) of a possible KGB agent or leaks from within A4. It was clear to all within the Security Service that Lonsdale was a very capable spy, well versed in counter-espionage techniques, and that he must not be alerted in any way to MI5's interest. There could be no mistakes. The strict new controls forbade any overt surveillance of Lonsdale, rigid radio silence was enforced, and a new sequential technique of surveillance was devised.

One team of watchers picked up Lonsdale when he left Wardour Street and followed him for a relatively brief distance before he vanished. The next day a new team picked up Lonsdale at the point where the previous team had retired, and followed their quarry for another segment of his journey westwards from central London. The same faces were never used twice, and so the wives of MI5 officers and volunteers were recruited to make up the numbers as the distance from Soho lengthened. On Monday, 24 October, Lonsdale's routine changed.

He travelled to Great Portland Street to collect his cases and deed box from the Midland Bank. The bank immediately alerted MI5 and a team of A4 watchers arrived within minutes to tail Lonsdale. He returned to Soho to deposit his possessions at his office, and took the Tube from Piccadilly Circus out to Ruislip Manor station, in the distant north-west suburbs of London. Here Lonsdale was followed along various roads. The watchers prudently kept at some distance because, as was his habit, he doubled back on several occasions to identify pursuers. Walking along Cornwall Road in Ruislip, he suddenly vanished down a narrow alleyway leading to a row of garages. Fearful of being spotted, the watchers did not dare tread on his heels. In the late afternoon on the Friday, the watchers trailed Lonsdale back out to Ruislip and tantalisingly further from the Tube station. In the gathering gloom, two women officers saw him bustle into a cul-de-sac called Willow Gardens. Here Lonsdale appeared to become very alert, glancing behind and around him nervously. Willow Gardens culminated in a narrow pedestrian passageway, where Lonsdale vanished from sight. Again the watchers deemed it too risky to follow, fearing he might conceal himself in the shadows at the end of the alleyway.

The passageway led into a street named Cranley Drive: a line of bungalows, many identical in design with sloping tiled roofs, bow windows, chimneypots and low privet hedges. No natural cover was offered on this open street and any observers sitting in a car for a lengthy period would be swiftly spotted, especially at a time when cars were still beyond the reach of many and residential roads were often devoid of them. To discover Lonsdale's destination, Elwell pressed for an OP to be set up in Cranley Drive by 5 November, ready for the possibility that after a meeting with Houghton, Lonsdale might go out to 'his lair at Ruislip'.[65] As usual in such cases, the Security Service asked the local police if they knew of a reliable family. The police superintendent recommended a couple called Bill and Ruth Search, and discreet security checks were carried out. Bill Search was an aeronautical engineer and travelled extensively, and his wife stayed at home looking after their two teenage children. The Searches lived in a house conveniently opposite the exit from the passageway leading into Cranley Drive, into which Lonsdale had vanished a few days before. Two windows on the first floor would enable MI5 observers to cover the length of the street. It appeared ideal. Jim Skardon arranged to meet Mrs Search at 4.30 p.m. on 4 November. As darkness fell, with streetlights starting to glow and curtains drawn tight to shut out the wintry gloom, he kept the appointment 'with some trepidation lest Lonsdale had actually been living at her address'. Ruth Search's first answers confirmed that no friends or relatives had been staying with her during the past month, so Skardon proceeded 'with greater confidence'. Although she took particular pride in 'seeing what is going on all around her', Mrs Search did not recognise the photograph of Lonsdale. She impressed Skardon with her intelligence and self-confidence. She agreed that he could use her address as an OP the next day and thought her husband, who would arrive home from Birmingham late that evening, would not object.

At the same time as Lonsdale returned to Britain, MI5 decided to intensify the surveillance of Houghton. While his cottage was empty one day, two men arrived at the front door. One of the pass keys clicked in the lock and within seconds they were inside. An eavesdropping device was installed in the telephone in the sitting room. Despite the sound being distorted and at times extremely difficult for the transcribers in the 'Gristery' to follow, the new surveillance started to bear fruit within days. On the evening of 22 October, Harry commented to Bunty:

'It's getting them back again, ain't it?
B: 'I couldn't do that.'
H: 'No, [? that's just] the thing . . .'
B: 'The only way I could get 'em would be if they come in our tray
by mistake like they do sometimes.'

The transcript was disjointed and scrappy, but some of the remarks
seemed to provide circumstantial evidence of Houghton's spying
and Gee's entanglement.

From a letter intercept the Security Service learnt that Houghton
planned to travel to London and stay at the Victory Ex-Services Club
in Seymour Street near Marble Arch on the night of Saturday, 5 November. It looked as though his next planned meeting with Lonsdale
was going to happen. On 31 October, a little before midnight, MI5
eavesdropped on a phone call to Houghton's home number from
a woman friend called Polly Pollard. She lived at an expensive and
fashionable address in Bloomsbury, and her acquaintance with the
Portland naval clerk piqued the Security Service's interest because
she had formerly worked at MI5, was 'very well connected' socially,
and probably moved 'in different social spheres [to Houghton]'.[66]
In the call Houghton confirmed he was coming up to London that
Saturday and would meet Polly then, not least to pass on some
juicy gossip about a mutual acquaintance who had shown him
'some old pornographic prints' which the man's wife knew nothing
about. Around the same time the bug in Houghton's home recorded
an incriminating conversation, which according to David Whyte
'virtually confirms our suspicions that Miss Gee is associated with
"Reverberate" in espionage and that she may be procuring papers
for him and bringing them out of the establishment in which she
works.'[67]

Whyte wished to keep his options open as Saturday, 5 November,
approached. It might be necessary to arrest Houghton, Gee and
Lonsdale that day. The MI5 investigators had already prepared a
list of people to be interviewed if Houghton was seized. They now
drafted a seven-page summary of the information amassed on the
navy clerk as background for Special Branch and did not paint a flat-
tering portrait: Houghton 'has a high opinion of himself and gives
the impression of trying to make friends with people rather above
his own social level. Besides his steady relationship with Miss Gee,

he appears to have flirtations from time to time with other middle-aged spinsters who, in spite of his coarse manner, evidently find something attractive in him.' Gee was 'in all probability active as a source of information for Houghton and probably removes papers for him from UDE'. As for Gee herself, 'Plain in appearance and speaking with a fairly strong Dorset accent, it would be hard to find someone further removed from the popular conception of the female spy than Miss "Bunty" Gee.' Although it might seem improbable from her outward appearance that Gee was a KGB agent, the Joint Intelligence Committee nonetheless allotted her a code-name. It had a resolutely horticultural air: 'Trellis'.

That week Roger Hollis was on leave so Graham Mitchell stood in for him. Mitchell was decisive: no attempt should be made to arrest Houghton that weekend because the Security Service needed to 'play it long'. Lonsdale would hopefully lead MI5 to other spies. When Mitchell attended the regular Thursday meeting of the Joint Intelligence Committee in Hollis's place, he was taken aside by the Director of Naval Intelligence, who said he would like to have a talk about Houghton soon after his 5 November meeting in London. Mitchell agreed. He explained bluntly that no arrest of Houghton was planned that weekend.

The A4 watchers made extensive preparations for the weekend surveillance operation. Whyte, Elwell and Skardon were determined that this time, unlike at the beginning of October, Houghton would not evade his pursuers. Whyte issued what he called a 'directive' to the A4 watchers confirming the new and stricter surveillance rules for Lonsdale:

> 'Last Act' has recently shown himself to be rather jumpy. We have therefore abandoned for the moment any attempt at continuous surveillance on him. He will almost certainly take evasive action before a meeting on Saturday. No attempt, therefore, should be made to follow 'Last Act' to the meeting on Saturday. An attempt should be made, however, to follow 'Reverberate' to the meeting place ... Wireless silence will be observed within a twenty-five-mile radius of London by A4 [watcher] teams bringing up 'Reverberate' from Dorset. Other A4 teams will observe wireless silence after the normal working hours on a Saturday, i.e. circa 15.00 hours.

North of Weymouth at a place called Culliford Tree stands a group of megalithic burial mounds. On a high point nearby, as a murky dawn broke to the east on Saturday, 5 November, the watchers parked their radio van, providing communications with vehicles over a twenty-mile radius. Nestled next to it was the Dorset police Land Rover, used by the Constabulary's dog handler, which was in direct radio contact with the local police network. This was emphatically analogue surveillance in an analogue age. Houghton was tailed by a succession of A4 cars to Ringwood and New Alresford on the way to the capital, before arriving in the south-west London suburb of Merton in the darkness of early evening. Here Houghton seemed lost, meandering in the backstreets, so a few officers risked leaving their cars to observe him on foot. There was a moment of panic when Houghton approached one of them asking for directions. The officer replied coolly that he had no idea, to which Houghton retorted, 'I'm buggered if I know either.' Houghton eventually found his way to the address of some old acquaintances and then drove to Surbiton, where he parked his car in a side street a little before 6.30 p.m. near to the Maypole public house.

At 6.27 p.m. Lonsdale appeared on foot on the opposite pavement carrying a briefcase. The pair met, walked to Houghton's car, and drove to another side street. Houghton parked in the darkest part of the road, switched off all the car lights, and the two men sat in the vehicle talking for about thirty minutes, the watchers keeping a respectful distance. The pair drove to the Maypole pub and stayed drinking there 'completely at ease' for about half an hour, the surveillance team noting that Lonsdale now held a slim, zip-fastened black document case rather than the briefcase with which he had arrived for the rendezvous. Houghton drove off with Lonsdale and the watchers lost contact with his car on the Kingston bypass.

During that Saturday, observation posts were also manned to cover both Lonsdale's flat at the White House and the Searches' house in Ruislip, where the Security Service stationed an officer upstairs in their daughter's bedroom. He saw nothing of interest during the day. After the meeting between Lonsdale and Houghton in the early evening, around 9 p.m. Jim Skardon suggested that the officer in Ruislip walk around the area. He saw nothing. Forty-five minutes later the watcher trudged round the same circuit, walking south from Cranley Drive through the narrow and unlit passageway where Lonsdale had previously disappeared. There, parked between two other vehicles, and unmistakable under the glimmer of a street lamp, was the bulk of Lonsdale's distinctive American car with its white wall tyres.

3

Code-name 'Killjoys'

I

On the Sunday morning Lonsdale's car was still parked in the same place in Ruislip, and a watcher manned the observation post in the Searches' house at 1 Courtfield Gardens. Jim Skardon decided to visit the area to oversee the next stage of surveillance. A4 assumed that Lonsdale was in a house close to where his car was parked and were looking for a second OP which had a direct view of his Studebaker. They had already been given the names of two people living in the block of flats next to the car who might be willing to help. Puffing on his pipe and wearing his trademark trilby, Skardon decided to call on the Searches to ask if they could suggest more names for a second OP. Skardon was impressed by Bill Search, finding him 'an extremely sensible, friendly and proper person', and 'without going into details . . . gave him an indication as to the reason for our presence in the Ruislip area'. Skardon asked if there were any 'foreigners' living in the area. Bill Search, after ruminating for a moment, mentioned one living opposite: a Canadian, Peter Kroger, an antiquarian bookseller who had been based in the area of the Strand in London but now worked from home. Numerous people called at his house and Search remembered that Kroger had lived in Catford, south-east London, before moving to Ruislip. Sensing that this information might be important, Skardon climbed upstairs and passed it on to the A4 officer based in the daughter's bedroom, suggesting he should focus attention on the Krogers' bungalow opposite.

During the late morning, Bill Search was cleaning his car outside his garage when Peter Kroger came across the road for a casual chat. Wearing horn-rimmed spectacles, he looked in his mid-forties: slim, with a sallow complexion, his square face was topped with grey-flecked white hair. A woman assumed to be Mrs Kroger appeared and drove a black Ford Consul out of the garage attached to their bungalow at 1.55 p.m. and returned indoors. The watcher had come downstairs with a coffee cup when he and Ruth Search saw the front door of 45 Cranley Drive open. The squat and unmistakable figure of Gordon Lonsdale emerged. It was a re-markable moment. Three weeks or so after Lonsdale's return to Britain, MI5 had finally tracked him down to what Elwell called his 'lair'.

Lonsdale glanced over his shoulder and sidled out of sight down the narrow passageway next to the Krogers' bungalow towards his car and disappeared, but without moving the Studebaker. Mrs Kroger drove off alone at 2.25 p.m. and returned home at 4 p.m. Three minutes later, Lonsdale reappeared from the passageway, cast a studied look behind him and hurried into the Krogers' house. He was not seen again that day.

In the comfort of their sitting room, Ruth and Bill Search sketched out more details of the Krogers for MI5. Both were about forty-five years old and had told their neighbours they were Canadian. Peter, the Searches observed, was quiet and friendly, and Helen pleasant and a 'good mixer'. They had no children. Ruth Search said that because she came from the New Cross area in south-east London, and the Krogers had mentioned that they had first lived in neighbouring Catford, she had often tried to talk about streets in the area but the Krogers were 'generally unwilling to discuss the Catford dis-trict'. Peter was reluctant to talk about his antiquarian book business in any detail. The Krogers had been away for four or five weeks on a business trip to Europe earlier that year. Although she found the Krogers friendly, Ruth Search had rarely invited them into her house. When the Searches had on occasion visited 45 Cranley Drive, they noted there was no television set. The watchers had already registered that there were no radio aerials on the roof. Later that Sunday evening the telephone rang in the hall of 1 Courtfield Gardens. It was Jim Skardon calling Bill Search to underline the need for 'the utmost discretion and to enjoin his wife to maintain a similar silence'. In the surveillance report he sent to Charles Elwell the next day, Skardon expressed confidence that the Searches would not be 'indiscreet' and their two children, a boy and girl, both at grammar school, 'will be sufficiently under control also to avoid any sort of public indiscretion'.

The following week the scent of the chase animated everyone

involved and the pace quickened. Charles Elwell began to investigate the Krogers.[68] Since they had said they were Canadian, he immediately asked the Royal Canadian Mounted Police for details. The RCMP had no record of them or any passports in their names. Their nationality was a mystery, although MI5 received some information that Helen Kroger was the child of English parents and forty-eight years old. Elwell did manage to confirm some details about the Krogers vouchsafed by the Searches. They seemed to have arrived in the UK in December 1954, when they first lived in the Catford area and Peter Kroger established his antiquarian book business. Soon afterwards they bought the bungalow at 45 Cranley Drive. Peter Kroger moved his business there in 1958. Within days of Kroger being identified by Elwell as a book dealer specialising in 'Americana and . . . fetters, handcuffs, leg-irons and instruments of torture', Elwell concluded it was likely that the Krogers might be the communications officers for Lonsdale's illegal network and could be hiding microdots in the books they sent abroad. The Post Office immediately started to photograph the packaging of all the books that Kroger despatched to Europe to harvest details of their addressees, mostly in Holland, and Elwell asked MI6 to seek information about them. A tap was placed on the Krogers' telephone and they were allocated the code-name of the 'Killjoys'. The Krogers were placed under surveillance by the watchers, but strict rules limited the use of radio for fear of interception by the Russians. Cars operating in the so-called 'Killjoy' area were to use a special frequency ('Lovebird'), but in 'no circumstances should . . . "Lovebird" . . . be used within half a mile of Kensington Palace Gardens [the Russian embassy]'. The FBI confirmed that the Krogers did not have US passports and they had no record of the Krogers having left America in December 1954 to travel to the UK.

On 10 November, the hidden microphone at Houghton's cottage eavesdropped on some new and incriminating material, despite the sometimes appalling audibility. Houghton and Bunty Gee had returned late that night and the bug picked up the following exchanges:

H: '[? When] you've been a bloody spy . . . it's horrible. It makes you feel as though you're—'

B: '. . . [? going to crack.] I definitely . . . well, you know, I got the flipping [? jonkers.]'

H: 'Yes. You've got the [? jonkers] every time you've come . . . But in London though er – Gordon is doing [? his bit.] . . . Yes, you work yourself up.'

B: 'I work myself up – I couldn't sleep.'

After Houghton's meeting with Lonsdale the previous weekend, and in light of all the accumulating evidence, it was clear that the case was of major importance and that Lonsdale and his network were being controlled by Moscow. Senior management at MI5 decided that the investigation should no longer remain with David Whyte but be shunted across to the Soviet section of counter-espionage, D1.[69] Around 14 November, direction passed from Whyte to the head of D1, Arthur Martin.

Martin's MI5 career had been turbulent. Recommended to MI5 in 1946 by no less than Kim Philby, in the summer of 1948 he was one of a tiny circle of Security Service officers who were indoctrinated into the work of American cryptographers deciphering top-secret Soviet wartime intelligence traffic – the 'Venona' decrypts. It was these decoded messages that provided vital clues during the investigation of KGB spies Klaus Fuchs (who confessed in 1950) and Donald Maclean (who defected in 1951). Martin had worked on both cases, together with Charles Elwell's wife, Ann Glass, who was still in the Security Service at that time. Although highly talented, Arthur Martin could be temperamental and tone-deaf to office politics. After the defection of Burgess and Maclean, Martin pressed unsuccessfully for an aggressive investigation into the network of KGB Cambridge infiltrations in the 1930s, but, frustrated by the lack of support, accepted a job in 1952 with MI5 in Malaya. In recognition of Martin's outstanding talents, Furnival Jones, after he was appointed head of counter-espionage in 1956, repatriated Martin and first appointed him head of Polish and Czech counter-espionage and then, from January 1960, D1.[70]

When the case moved to D1, so did Charles Elwell. Elwell was to run the continuing investigation, helped by another officer, William ('Bill') Colfer.[71] There is no evidence of the reaction of either David Whyte or Charles Elwell to this brusque decision. Elwell was a loyal and professional intelligence officer. Although he did not perhaps cleave as naturally to Martin (a grammar-school boy who had been an army Signals officer in the Radio Security Service during the war) as he did to his friend David Whyte (a fellow former public-schoolboy who saw active service during the war), Elwell respected Martin's expertise in Soviet espionage and set to work with his customary zeal. After the Portland investigation was shunted over to the Soviet section, except for isolated mentions David Whyte disappeared completely from the MI5 files.

Meanwhile, MI5 senior management skirmished with the Admiralty, which was increasingly apprehensive that MI5 were allowing Houghton and Gee to continue to smuggle secret information out of Portland without hindrance. The Admiralty demanded to know when the investigation would end. The Director of Naval Intelligence, Admiral Denning, forthrightly posed this question to Graham Mitchell when he visited him on 16 November. In response Mitchell havered: he did not know, and if MI5 'continued to derive secret intelligence from [its] investigation of "Reverberate" and those associated with him, it might last quite a long time'. The Security Service now knew definitely from the GCHQ intercepts of Lonsdale's radio traffic that Russian espionage was involved, and hoped that he would lead them to other spies, although the early decrypts concerned only 'Shah' and 'Asya', the KGB code-names for Houghton and Gee, and family news about Lonsdale's wife and children.[72] One reason for the Admiralty's trepidation was revealed during the meeting. UDE was about to start highly secret work on a new type of torpedo. Although neither Houghton nor Gee would have authorised access to this project, Denning 'was worried lest things should get about'. He confirmed to Mitchell that the government minister responsible for the navy, the First Lord of the Admiralty, Lord Carrington, had been told about the Portland case but said there was no risk he would pass the information on to anyone else, including Cabinet colleagues.

The bug in Houghton's cottage provided further titbits of information about his deceit towards Ethel Gee and, more importantly, of their spying. For example, on the evening of 17 November Houghton picked up a work colleague called Sylvia Brown (who was in the teleprinters' department at UDE and, like Houghton, was evidently fond of a tipple) and drove her out to his house at

9 p.m. No conversation was detected until she was about to leave just before midnight when, according to the MI5 transcript, there were 'further giggles from the woman because Harry was kissing her and the curtains were not drawn. Harry said it would give the b-----s a treat.' Three nights later, when discussing Christmas with Gee, Houghton 'speculated as to what "he" [presumably Lonsdale] might buy them [as a present]'. Houghton went on: 'The only thing is, got a washing machine and an electric razor, thanks to him.' Bunty was giggling all the time, and said: 'We didn't do so bad last Christmas.'

'Dust Cover' was an apt code-name for the observation post at 1 Courtfield Gardens in Ruislip. The area was steadfastly suburban, anonymous and middle-class: 'one of those places one drives through on the way to Oxford . . . neat rows of semi-detached houses; small front gardens, each with a square of lawn and herbaceous border; bay windows; clipped hedges; and every so often, where the down-stairs curtains have yet to be drawn, the bluish flickering light of a television set'.[73]

Although Ruislip's history dated back to the Domesday Book (where it appears as Rislepe, thought to mean 'leaping place on the river where rushes grow'), it had only burgeoned into a sprawling suburb after the First World War, when King's College, Cambridge, sold land for development to house commuters using the newly extended Metropolitan Line. Almost without exception, sofas and armchairs were shrouded in dust covers and antimacassars by house-wives of the immediate post-war generation for reasons of thrift and respectability. For the two adolescent Search children, stirred by the distant drums of 1960s teenage rebellion, Ruislip was synonymous with boredom – *nothing* ever happened there. Yet, with respect for authority still almost universal among the young, there was no question of fifteen-year-old Gay Search and her elder brother disclosing to their friends the secret of mysterious strangers keeping watch on neighbours from the family home.

These strangers, MI5 decided, should be a relay of two young female officers (a recent innovation opposed by Jim Skardon, partly through fear of inappropriate sexual dalliances between the nubile new recruits and his existing male stalwarts). One female officer, rather daringly for the time, arrived alone on a motor scooter. Both entered the house every day by the back door and kitchen to avoid attracting attention. By turns they lurked in Gay Search's bedroom (where a special telephone line was installed) with a pair of binoculars, logging movements to and from the Krogers' bungalow opposite. The requirements of espionage had on occasion to give way to prudence or the needs of domestic routine: A4 had to suspend observation when the Searches had visitors or Ruth Search needed exclusive use of the kitchen.

Despite all MI5's precautions and the Searches' discretion, there was always a risk of detection. One day, for example, Ruth Search was in the kitchen with the two female officers when they heard the back gate rattle and saw Helen Kroger approaching the house. The two watchers scrambled into the downstairs bathroom. Her heart pounding, Mrs Search opened the door and let Helen in. They started to chat when Ruth Search, with a shudder, spotted that one of the two women in her panic had left her handbag on the floor. Thinking quickly and realising the handbag required explanation, she blurted out to Helen: 'Oh that daughter of mine, she never puts anything away!'[74]

As the days passed, however, the inconvenience caused by the OP to the Searches, and especially Ruth Search, soured to 'embarrassment' and 'some dismay', as Skardon termed it in a note to Elwell on 24 November. Skardon visited Mrs Search that day to reassure her. She revealed that she had become 'a bundle of nerves', caused mostly 'by the fact that she has a number of friends living locally who call upon her from time to time'. Already the side garden door had been locked to prevent the risk of Helen Kroger using it again unexpectedly in case she might 'look through the curtains and possibly spot our officers in the kitchen'. Ruth Search asked to be able to tell one of her best friends about the observers being based in the house. Skardon was sympathetic and asked Elwell if Mrs Search might inform the friend that officers were stationed in the house on 'a police matter'. If MI5 did not agree, there was a grave risk that the Searches would insist that the OP in their house must be closed down. The watchers had quartered the area, but there was no alternative site for an observation post.

Elwell understood the threat this posed to the whole investigation. The idea of telling a close friend of Ruth Search was highly risky. If one friend knew, why not another? Elwell boldly decided to contact Bill Search himself and arranged to meet him in a local pub. Set snugly in a corner, Bill Search straightened his brow-line spectacles, ran a hand through his receding grey hair, and whispered his and his wife's apprehensions about the OP continuing. Elwell focused his attention on him like a searchlight. He was an excellent listener when occasion demanded. He appreciated with growing dismay that the Searches were serious. There was a genuine risk that MI5 might be forced to abandon the surveillance at 1 Courtfield Gardens.[75]

After talking to Bill Search for several minutes, Elwell realised he needed to take a fateful decision. To persuade the Searches to allow the OP to continue, Elwell judged that he would need to initiate him into some of the secrets of the case, to make clear what was at stake. Could he take the immense risk of trusting the man sitting in front of him? All his years of MI5 experience urged caution. After a pause of several seconds, he decided to embrace the risk. The OP was too important to be abandoned at this crucial point in the investigation. He leant forward, reminded Search of the draconian provisions of the Official Secrets Act, and in a lowered voice revealed much more to Bill Search than any prudent case officer would normally have dared. Search listened warily at first but behind his spectacles his eyes widened. Elwell finished his explanation. Bill Search agreed that 'Dust Cover' could continue but none of their friends be told.[76]

II

The Searches were not especially close friends of Helen and Peter Kroger, more good neighbourhood acquaintances.[77] Helen Kroger, with her characteristic gap between her front teeth, was an extrovert and on good terms with many residents, including the Searches, on whom she called most weekdays but not at weekends. She was renowned in the area for her boisterous manner, stentorian voice, and, to shock the staid burghers of Ruislip, instead of always wearing a dress or skirt she regularly sported slacks. The Searches' daughter, Gay, remembered Helen once whistling all the way down Cranley Drive through two fingers. By contrast Peter Kroger struck neighbours as a quiet, bookish man. He rarely entered the Searches' house. From the upstairs window of 1 Courtfield Gardens the female officers observed routine arrivals and departures: the postman, the milkman, deliveries of books for Peter Kroger and occasional callers connected to the antiquarian book trade. The only person of interest was Gordon Lonsdale, who visited on a number of weekends in November and December.

In common with most residents of Ruislip in 1960, the Searches did not have central heating. To relieve the biting winter cold and the tedium, the women watchers would periodically descend to the kitchen and chat with Ruth Search. After MI5 had revealed to her and her husband that the Krogers were more than simply the antiquarian bookseller and his wife they represented themselves to be, Ruth Search became entangled in an extraordinary game of pretence with Helen Kroger. As Gay Search later recalled, it 'became a game of wits. Mum knew Helen had lied to her and was still lying. I think Mum's attitude was, well, two can play at that game.' As the days passed the tension grew, and sometimes, when her children tried to relieve it by playing the fool, Ruth Search flared into an uncharacteristic bout of anger. Once in the kitchen her son decided to do a 'Russian dance, kicking his legs up. He thought it was hilarious but Mum really lost it.'

The surveillance continued throughout November, drawing closer and receding like rising and falling tides as MI5 observed signs of suspicion or watchfulness on the part of Lonsdale and the Krogers, perhaps caused only by them seeing 'at times only shadows' (Skardon's phrase in one memo). The bug on the Krogers' telephone picked up their apprehension over evidence that someone had tried to force their front door lock and their worries about men entering the loft when doing repair work at the bungalow (Elwell speculated that 'the loft may contain some equipment connected with espionage', perhaps 'a wireless transmitter'). Lonsdale not only stayed overnight with the Krogers at weekends but met them on one occasion in central London. On 23 November, Peter Kroger was observed walking to Selfridges department store in Oxford Street and stood under its large clock. At 6 p.m. precisely Lonsdale approached him, followed by Helen Kroger, who kissed them both. They stood chatting for a few minutes – Lonsdale appearing 'slightly on edge and looking about to see that all was well' – before they retreated to a local pub, later driving off together in the Krogers' car.

The GCHQ intercepts of Lonsdale's radio traffic to and from

Moscow consisted only of messages about how to handle Hough-ton and Gee and about his family, but contained no intelligence derived from Portland. Elwell and his colleagues assumed that the Krogers were providing this vital conduit of communication for the KGB. Based on evidence of how Soviet illegals operated, the Se-curity Service supposed that the Krogers were acting as Lonsdale's radio operators and also transmitting stolen, secret information by microdots and servicing dead drops. By the late 1950s GCHQ had developed technology to locate and intercept radio signals transmitted by KGB illegals in the UK, but in order to use it they needed MI5 to tip them off about the approximate time and place of transmission (to a mile or so). It did not have the capability or capacity to scan large areas of the UK on a permanent basis to pick up potential radio traffic generated by illegals. This started to change in 1959, when the Interdepartmental Policy Committee, which decided all GCHQ's priorities, took a top-secret decision that, from the end of that year, Cheltenham should start to create such a capability so that it could act independently and provide leads to MI5. Before then the Security Service and GCHQ did not believe that Russian agents in the UK were likely to use radio trans-mitters because they had other, more secure means of communica-tion, such as diplomatic bags or microdots. The change of policy resulted from growing evidence that the Russians had developed increasingly sophisticated 'flash transmission' technology which compressed the sending of a lengthy signal into a few seconds, a shooting star of radio traffic which vanished almost as soon as a broadcast began, and portable equipment small enough to use. Moscow considered this equipment secure enough to deploy in the UK. By the end of 1960, however, this new GCHQ project was only in its infancy, hampered partly by its inability to recruit sufficient skilled staff.[78]

After some delicate negotiations, MI5 and GCHQ agreed that Cheltenham would be allowed to make an attempt to intercept trans-missions from 45 Cranley Drive. This would be done by despatching a specialist detector van to Ruislip in December. These vehicles had been developed in secret, some disguised as Post Office or delivery vans with plastic bodywork to keep interference to a minimum, and were crammed with specialist equipment. A date for the experiment was fixed. The GCHQ team were carefully briefed by MI5 about the operation to minimise the risk: Arthur Martin and Elwell were terrified that the sight of an unfamiliar vehicle driving to and fro in Cranley Drive would arouse the Krogers' suspicions. No one knew the wavelengths on which any radio in the bungalow would transmit, so for this reason it was crucial that the detection equipment was deployed very close to the bungalow. On the appointed day, the van would have trundled slowly along Cranley Drive and halted a couple of times, attempting to appear inconspicuous. It would not have been permitted to make several passes along the street, as GCHQ would normally have expected, for fear of alerting the Krogers. The vehicle looped around the surrounding streets at various distances from the Krogers' bungalow, the technicians crouched over their equipment, alert to the faintest bat-squeak of a transmission in their headsets. They found nothing. There was some half-hearted discussion between GCHQ and Leconfield House about repeating the operation. MI5 vetoed the idea. The threat of alerting the Krogers was too great when balanced against the chance of success. The odds of being in the area at the precise moment the Krogers were transmitting a signal and intercepting it were infinitesimally tiny. Arthur Bonsall of GCHQ later described the operation as 'something like the needle and the haystack . . . [The chance of success was] something like 100,000 to 1.'[79]

The Krogers had installed new and secure locks on the doors and windows. Neighbours provided sketchy descriptions of the interior – books lining the walls and a large radio in the sitting room – but, understandably, if the Krogers were professional spies, there was no visible indication of any transmission equipment, either inside or on the roof of the bungalow. It was much too perilous for MI5 to burgle the Krogers' redoubt in the way it had entered Lonsdale's flat in the White House. So the inside of the bungalow remained mysterious, like the inner sanctum of a temple to which access was forbidden.

The Krogers also retained their mystery. Elwell painstakingly assembled the known fragments of their lives like pieces of a jig-saw: arriving in London on Christmas Day 1954; renting a house in Catford before buying 45 Cranley Drive in December 1955; Peter Kroger establishing himself as an antiquarian bookseller, first in the Strand in August 1955 and then operating from Ruislip after December 1958, providing generous hospitality to fellow booksellers and neighbours. The enquiries progressed on tiptoe for fear of alerting the Krogers. One of the strangest facts to emerge, which took several weeks to establish, was that although the Krogers told neighbours they were Canadian (they bridled when some suggested their accents were American), the residents of 45 Cranley Drive did not have Canadian passports but New Zealand ones, issued in April 1954 by the New Zealand legation in Paris. They spoke of their lives in Canada before 1954 only sketchily, saying that they had lived 'in the backwoods' there, before steering the conversation to other topics. The Krogers appeared in England from misty blankness at Christmas 1954.

Interceptions of Houghton's phone calls at home and work revealed that behind Bunty Gee's back he was continuing to meet Sylvia Brown.[80] The new D1 case officer, 'Bill' Colfer, travelled down to Dorset to meet the senior staff at Portland and the police in Dorchester in late November as part of the handover from D2. He stressed that MI5 wanted to run the case on for as long as possible and nothing should be done to alert Houghton and Gee. Boggust was asked 'to lay off his surveillance', but MI5 gave the special sergeant a present for his efforts: £25 in cash. Through his father's neighbourly snooping, that Christmas the junior Boggust would have reason to celebrate.

Eavesdropping on Houghton and Lonsdale suggested that the clerk was planning to meet Lonsdale in London on Saturday, 10 December. Houghton's forthcoming trip to London prompted a troubling question for the Security Service. From all the surveillance operations and interception of Lonsdale's radio traffic with Moscow so far, somewhat to MI5's surprise and perhaps disappointment, in the deciphered messages there were still no references to any additional assets that Lonsdale might be controlling. The only agents mentioned were 'Shah' and 'Asya', Houghton and Gee. It was therefore planned tentatively to arrest them in February or March 1961 while they were passing confidential material to Lonsdale. But what if, Elwell pondered, the documents they had handed over proved harmless, were perhaps not even official papers?

Elwell shared his worry on 7 December first with Arthur Martin and then with the head of counter-espionage. Martin Furnival Jones directed Elwell to consult the Admiralty to 'see what could be done to ensure that ... some fresh confidential information would be naturally introduced into the drawing office to which "Trellis" would have access, in the normal course of her duties'. In fact, nothing further did need to be done. This was because, based on an intercepted conversation of Houghton's, MI5 learnt (no doubt with some amazement) that Portland was operating a new security system and Houghton said he now had the key to the courier's box for the naval centre – the box used to move sensitive documents from user to user around the base. Elwell and Colfer shared this information with Naval Intelligence on 15 December. After discussion, the new DDNI took the bold decision not to pass this information about the courier's box on to the Captain in charge of UDE 'for fear of his taking steps to prevent Houghton's access to it. DDNI thought that access to the courier's box was the greatest threat.'

Elwell began Saturday, 10 December, in buoyant mood in the Security Service operations room at Leconfield House. As the morning unfolded, however, the tension heightened. The Dorset police keeping watch on the main roads to Salisbury did not spot Houghton's car, and the A4 officers waiting at Salisbury railway station did not see Houghton or Gee arrive to take a train to London. Attention switched immediately to Lonsdale's flat near Regent's Park. The suspected Russian agent left the White House at 3.05 p.m. and was observed by A4 to arrive and drive around the Waterloo area before parking and walking towards the Old Vic carrying a 'white carrier bag of the type which is supplied by shops which sell ladies' lingerie, and this seemed well filled' – undoubtedly, as Elwell remarked a few days later, 'with "Reverberate's" and "Trellis's" Christmas presents'. Not seeing Houghton, Lonsdale sauntered into Waterloo station and lingered by platform 14 for a few minutes before walking away to sit in the main booking hall. At 4.15 p.m. the watchers finally saw Houghton in the station, alone and strolling by the main departure board towards the entrance to the Underground, where he met Ethel Gee. They left the station and walked over to the Old Vic, where they finally joined up with Lonsdale – still with his carrier bag – and went into a restaurant opposite Waterloo station. As instructed, the watchers withdrew.

Meanwhile, at 1 Courtfield Gardens in Ruislip, the A4 female officers passed the day observing the Krogers' bungalow: a delivery of parcels in the late morning was taken in by Peter Kroger, and the pair made a couple of car journeys in the afternoon. After darkness fell and the sparse street lights began to glow, the female officer on duty started her periodic patrols around the streets neighbouring 45 Cranley Drive. At 6.55 p.m. she saw nothing of interest. But just over an hour and a half later, she noted the familiar bulk of Lonsdale's Studebaker, ULA 61. Elwell retired to bed that night with an inner smile of self-satisfaction: after an anxious start, the day had ended as he had planned. The next day did not, however, begin as hoped. The woman watcher duly appeared on her motor scooter at the Searches' house at 10 a.m. and climbed the stairs to the daughter's bedroom to start another wearisome day at the observation post. A4 had by now also set up a second OP which overlooked the area where Lonsdale parked his car. This was manned from the same time. At 10.45 a.m. the watcher in 1 Courtfield Gardens was slightly dismayed to observe Helen Kroger drive up to 45 Cranley Drive in their black Ford Consul with her husband and Lonsdale. They had clearly left earlier, perhaps around 9 a.m. Elwell was convinced that 'this early-morning expedition had some connection with the meeting between "Last Act" and "Reverberate" on the previous day'. Lonsdale scurried into the bungalow, leaving the Krogers to park the car in their garage. He was not seen outside 45 Cranley Drive again that day.[81]

MI5 searched Lonsdale's flat for a third time on 17 December.[82] Winterborn made detailed notes on the cipher pads for GCHQ, confirming how many pages and number blocks Lonsdale had used. A fruitless attempt was made to open the Chinese scroll while it was still hanging on the wall. In Ruislip, Helen Kroger paid Ruth Search a Christmas visit with presents for the children, and revealed that 'they were having Peter's American friend staying with them' the weekend before the festive break. Whenever Lonsdale passed a weekend at Ruislip in December the various OPs in the eight-roads area and Ruislip were activated to monitor his movements in case he led the Security Service to a dead drop site or clandestine meeting. Nothing of importance, however, was seen then or when Lonsdale visited the Krogers just after Christmas. A letter from Lonsdale to 'his lady-love, Denise', as Elwell termed her, was intercepted which disclosed his plans to visit her in Brussels sometime after 10 January 1961. Otherwise, for Elwell and D1, 1960 ended tranquilly. The

Portland counter-espionage investigation was under control and a provisional plan was in place to arrest the spy ring in February or March 1961.

That calm was to be brusquely shattered in the first few days of the New Year.

4

Code-name 'Sniper'

I

As 1961 began, 932 miles due east from London there was barely suppressed excitement in the CIA's Berlin Operations Base (known as BOB). The Berlin Wall had not yet been built, but the closure of the sector borders within and around the German capital seemed increasingly inevitable. The mysterious agent code-named 'Sniper' had continued to send letters to the CIA containing highly valuable intelligence. But as 1960 ended, the spy had hinted that he might soon need to defect to the West. An emergency telephone number was established on the BOB switchboard exclusively for 'Sniper's use. First a call came through on behalf of a Herr Kowalski, asking to set up a dead drop in Warsaw for 'Sniper'. A drop was made and the information recovered. The asset contacted BOB just after the start of the New Year and said he planned to defect very shortly. Careful preparations were made. It was crucial that the CIA reception team could establish their credentials with him quickly, so the American consulate was selected as the location of the momentous first meeting with the spy: it was open to civilians and close to the military section of the Clayallee American compound guarded by military police. No one knew either 'Sniper's appearance or identity.[83]

The CIA in Washington sent a coded cable to its London station and, on the afternoon of 3 January, it delivered by hand a sealed message to Roger Hollis in Curzon Street that 'Sniper' was on the verge of defection.[84] Alarm bells instantly rang at Leconfield House. If the agent defected there was, in Elwell's words, 'a serious risk that "Reverberate" and, therefore, "Last Act" would be compromised'. With senses heightened, MI5 watched for any sign that Moscow might warn Lonsdale. On New Year's Day GCHQ had intercepted an incoming message from the Moscow area to Lonsdale, but this predated the latest, electrifying intelligence from Berlin. The message had alerted Lonsdale to the imminent arrival of a communication from Moscow Centre containing three microdots, and Elwell made urgent arrangements to intercept it either at the White House or in Ruislip if it arrived by post.

On the afternoon of 3 January, the listening device in Lonsdale's flat detected that he was typing furiously, and just after 4 p.m. he tuned his radio to a programme in Russian for around fifteen minutes before switching the channel brusquely to popular music. Around 8 p.m. there was the sound, according to the eavesdropping device, of a 'Machine in use. (? camera)'. Lonsdale retired to bed early, flushed his toilet 'no less than nine times between 00.05 and 06.43', rose around 3 a.m. and moved around the flat for about two and half hours before getting dressed 'by 06.15, when he received a . . . signal [from Moscow]'. Elwell was apprehensive. 'This unprecedented behaviour by "Last Act", coupled with the possibility that the Russians may know that ['Sniper'] has been in touch with the Americans, suggested that "Last Act" might have been alerted and might be taking steps to leave the country.' Urgent orders were sent to London airport to arrest Lonsdale and the Krogers if they surfaced there ready to board a flight, and A4 crews were put on alert to tail them in case they fled elsewhere. An alternative explanation for Lonsdale's behaviour, Elwell mused, was that he 'was making another attempt' to send a message to Moscow. GCHQ were asked to try to intercept any signal sent that day from Ruislip between 12.40 and 15.20 – the times when the Security Service thought Lonsdale might have transmitted from there the previous Wednesday. All MI5 could do was continue to watch and wait.

Over in Berlin, on 4 January at 5.30 p.m. came another call from the Kowalski go-between, declaring that Kowalski would arrive at the consulate in thirty minutes and requesting that Mrs Kowalski be treated with special care. With adrenalin pumping, the CIA reception team moved to their positions: one officer each to the main and side entrances; one to act as 'roving security officer' to provide any necessary explanations to consulate employees; while another, who had flown in to West Berlin to act as Washington's 'special representative', walked to a grand office in the consulate, Room 3025, which had been specially fitted with microphones. Two other cars were parked at strategic locations outside. At 6.06 p.m. a West Berlin taxi pulled up outside the consulate and a man and a woman clambered out, each clutching one bag and the man a briefcase, muffled against the cold. The man sported an extravagant and drooping moustache, the woman an expensive fur coat.

Inside, the senior CIA officer announced that the couple would be offered asylum if the man identified himself and consented to be debriefed by the American authorities. There was a discomforting pause. Eventually the man, in a lowered voice in German, responded that the woman was not his wife but his mistress and requested asylum for her. He asked if the woman could wait outside while he discussed some matters of great sensitivity. The woman was escorted out and the man revealed that she did not know his true identity. She knew him only as a Polish journalist under a different name, whom she had met during one of his business trips to East Berlin, fallen in love with and, a top-secret CIA memo recorded, 'whom she had joined that day in an adventure destined to make it possible for them at long last to live as man and wife in the free West. Subject [the defector] repeatedly emphasised that news of what was really involved would have to be broken to Irmgard [the woman] very slowly, lest she suffer irreparable psychological damage.'

Asked to confirm his true identity, the man first produced his ID card but after a minute or two of CIA note-taking said that 'this was a waste of time' because the card only gave his cover name and profession. He cut open the lining of his briefcase and scrabbled around inside before finally producing a document identifying him as Lt-Col. Michał Goleniewski, a senior official in Polish intelligence who had written the 'Sniper' letters. Throughout Goleniewski kept murmuring that 'despite his presence in the US consulate he was still in considerable danger – did we know about the [redacted] murder and did we know that the Poles knew exactly where [redacted] was in the U[nited] S[tates]?'. When the couple were reunited at 6.35 p.m., Goleniewski revealed to his mistress for the first time that they were destined for eventual resettlement in the United States. 'Irmgard', the CIA memo noted, 'accepted this disclosure with the same stoicism with which she went through the rest of what, by any criterion, must have been the most surprising evening of her life . . . Subject stroked her hair and kissed her – gestures which were to be repeated several times during the evening.'

The jubilant CIA team immediately despatched a top-secret cable from Berlin to Washington confirming the defection. Goleniewski and his mistress were escorted to a CIA safe apartment in the American housing area two minutes' drive from the consulate and whisked away at first light the next day to a military plane, to be flown to the American airbase at Wiesbaden in West Germany. There 'Sniper' met for the first time the CIA desk officer in Washington, Howard Roman, who had been handling his case, and they were all flown immediately to the United States. Goleniewski told the CIA he would not be missed before the evening of 5 January.[85]

A second sealed message with news of the defection was hand-delivered to Leconfield House hours later. Roger Hollis convened an urgent meeting in his mahogany-panelled office on the morning of 5 January. Although an introvert by nature, he was instinctively collegiate, especially when risky decisions needed to be made. That Thursday morning the Deputy Director General of MI5, Graham Mitchell, the head of counter-espionage, Martin Furnival Jones, Arthur Martin, David Whyte (this is the last time Whyte was formally involved with the investigation), Elwell and MI5's legal adviser all sat round Hollis's meeting desk. Furnival Jones outlined the defection, Goleniewski's career in the Polish Intelligence Service and his conclusion that 'sooner or later the Russians would deduce that he [Goleniewski] had known about' Houghton, and as a result might exfiltrate Lonsdale and the Krogers. He feared that the Russians might learn about the defection as early as that evening and warn Lonsdale. The Security Service knew from GCHQ's monitoring of Lonsdale's radio traffic that he was expected to receive another message from Moscow early in the morning on Saturday, 7 January, and it seemed likely that if the Russians wished to alert their illegal they would use this channel to do so. The Security Service understood that Houghton was planning to meet Lonsdale in London on 7 January.

The six men discussed the various options. Lonsdale was planning to visit his mistress in Brussels for a few days after 10 January. If MI5 delayed his arrest he might never return to the UK. There was no doubt that the best moment to arrest Houghton and Lonsdale was when they were meeting. Furnival Jones was clear in his recommendation: MI5 should plan to arrest all five suspected spies that Saturday, in two days' time. Hollis agreed. It was, as Hollis characterised it a few months later, 'a jolly hard decision to make'. Goleniewski's defection 'forced' MI5's hand: Lonsdale and the Krogers needed to be 'grabbed . . . before they were able to get out'.[86] Ever mindful of Whitehall protocol, at the meeting of the Joint Intelligence Committee the same day Hollis alerted the Director of Naval Intelligence to the likely arrests, and on the Friday evening he went to see the head civil servant at the Home Office, Sir Charles Cunningham, to warn him also. Hollis did not think the news would break until Monday morning, when the suspects appeared in court. The Home Secretary was at his country house until the following Thursday, but Cunningham knew that Rab Butler would understandably be 'annoyed' if he read about the arrests first in the newspapers, so the Permanent Under Secretary sent him a note 'in his own hand' in the ministerial box at noon the next day. Cunningham was very relaxed about informing the Prime Minister: it would be sufficient for MI5 to tell Number Ten on Monday morning before there was any publicity. This was symbolic of Macmillan's relative lack of interest in intelligence matters, and of widespread unawareness of the political repercussions for the government of this espionage case. The Portland spy ring was, in fact, to remain a political blind spot for Macmillan for several months.

The following thirty-six hours passed in a whir for Elwell and Colfer as they made frantic preparations for the arrests. Elwell's detailed plan was refined at a meeting at noon on the Friday in Arthur Martin's room in Leconfield House. The two senior Special Branch officers, Superintendent Smith and his assistant, Chief Inspector Ferguson Smith, who were to make the arrests, attended, together with Peter Wright from A2 and Jim Skardon. The trap was set.

II

If the Britain on which the sun rose on the morning of Saturday, 7 January 1961, superficially appeared little changed from the same country five years before, deep down the cultural plates were shifting. At the end of December 1960 the Beatles, just home from their first brief exile in Hamburg, gave a sensational performance in the ballroom of Litherland Town Hall near Mersey docks. According to one observer, 'the audience went *mad* . . . That was the beginning of Beatlemania.' Magnificent if sometimes derelict Victorian buildings were being bulldozed across the country in the name of modernity, from Basingstoke to Birmingham, from Cardiff to Glasgow. The pace of change in the urban built environment was in some ways unmatched since the Industrial Revolution. Immigration from the New Commonwealth was starting to accelerate, jumping from 21,600 in 1959 to 57,700 in 1960. And on television, cultural icons of protean longevity were being created. The legendary soap opera of Northern working-class life, *Coronation Street*, began broadcasting as December 1960 drew to a close (the critic of the *Manchester Evening News* admiring the 'feasible characters and such convincingly ungrammatical and slovenly speech'). On 7 January 1961 *The Avengers* began, but then the main role was taken by Ian Hendry, and his assistant John Steed played by a fresh-faced Patrick Macnee. The early episodes presaged no hint of a Cathy Gale, and certainly not Emma Peel.[87]

The sky was clear that morning, stars still glittering through the frosty air, when at 6.30 a.m. a light flicked on in the upstairs bedroom of Houghton's cottage at the end of Meadow View Road. Beyond were fields, still wreathed in early-morning darkness. Minutes later a car eased to a halt on the main road and the headlights were doused. The driver waited inside while his colleague, muffled against the biting chill, sauntered up the road towards the cottage before scurrying back. The driver picked up a microphone in the car and reported that lights were showing in an upstairs room of Houghton's cottage.[88]

The car was one of five vehicles sent down to Portland the day before by MI5's watchers. The message was transmitted to their mobile radio unit stationed several miles away at Culliford Tree. At 7.50 a.m., Houghton – easily recognisable because of his wisps of hair and aquiline nose – emerged from the cottage. He scraped the frost from the windscreen of his car and lowered himself down into the driver's seat. The surveillance car tailed Houghton at a discreet distance as he drove towards the narrow isthmus which linked the mainland to the Isle of Portland. At 8.30 a.m. he was seen travelling back with Ethel Gee, dressed that day in a navy-blue suit.

The couple were followed as the navy clerk drove, cagily because of the ice gripping the roads, from Portland to the railway station in Salisbury. A London train was just about to leave. While Houghton parked his car, Gee bought two return tickets and they hurried through the barrier, being (as the MI5 surveillance report noted) 'urged on by a porter, but unfortunately, although "Trellis" covered the ground, "Reverberate" found the task too much for him and they just missed the train'. Houghton and Gee whiled away the next two and a half hours meandering around the shops and market in Salisbury until the next London train. The Atlantic-class steam locomotive puffed its way out of the station at 12.32, but staff were already announcing that its arrival in the capital would be delayed because of a landslide on the track ahead. On the train with Houghton and Gee, but unnoticed by them, were a handful of MI5 watchers and Metropolitan Police officers. One of them was Chief Inspector Ferguson Smith of Special Branch, a mustachioed, modest and highly intelligent officer who had been decorated for his bravery while serving in Bomber Command during the war. The police called ahead to London to warn which train Houghton and Gee had caught.

By contrast, Ferguson Smith's immediate boss in Special Branch, Superintendent George Gordon Smith, was a self-publicist. The superintendent was in titular charge that day of what the police called 'Operation Whisper'. He had passed the morning in his office in New Scotland Yard liaising with MI5, waiting for the trigger to be pulled by the Security Service for the next phase of the operation. Elwell had hardly slept in the previous two days as he frenziedly worked with colleagues in MI5, Special Branch and GCHQ to plan the arrests. With his directness and attention to detail, Elwell was a natural leader. Dressed in his customary pinstripe suit and bowler hat, he had as usual walked to work that morning. Before dawn the trees around Hyde Park stood leafless and stark in the wintry light from the streetlamps as Elwell hastened past – he needed to arrive early in the Security Service operations room on the third floor.

The room was spartan, resembling a prison cell, the walls painted a caramel-brown, with a bed pushed against one wall and a table in the centre. A tangle of cables trailed across the floor carrying the links to Special Branch, GCHQ and Roger Hollis, the Director General of MI5. A speaker was connected to the bug the Security Service had concealed in Lonsdale's rented flat. There were four men in the operations room when Elwell entered: his immediate boss, Arthur Martin, chain-smoking as usual; Martin Furnival Jones, sitting on the bed in braces; and the two technical experts, Hugh

Winterborn and Peter Wright, who had secreted the eavesdropping device in Lonsdale's apartment. Elwell arrived to hear the sound of murmured voices relayed from the bug. A woman had spent the night with Lonsdale and now he was persuading her to leave because he had urgent business to attend to.[89] Everyone in the room understood what that business was: to listen on his radio around 7.30 a.m. to the bi-weekly encrypted message from KGB headquarters at Moscow Centre. A GCHQ technician had travelled up to London from Cheltenham the day before expressly to transform the random digits into language the instant the message came through. MI5 were apprehensive in case their work had been compromised or betrayed and Moscow was about to warn Lonsdale.

Overnight two men had slept in the flat next to Lonsdale's: an MI5 officer from A4 and a Special Branch inspector, who would arrest Lonsdale if he attempted to flee. Other officers were stationed outside the block. Once Lonsdale's lover had departed, the operations room heard the sounds of the agent pulling out his radio set and preparing his pads to receive the message. There was silence as they listened to the crackle of Lonsdale's radio and the faint scratch of his pencil as he transcribed the decrypt. He suddenly began singing chirpily in Russian and the volume faded as he moved into the bathroom. The operations room relaxed and remarks were exchanged: surely Lonsdale would not behave like that if Moscow had raised the alarm? Although the tension dissipated, they waited. Eventually the green telephone rang carrying the scrambled line from GCHQ. The message from Moscow was standard: only brief news from home of Lonsdale's family.

Elwell and Hugh Winterborn left Arthur Martin in Leconfield House and drove to the A4 observation post which overlooked the White House. The MI5 flat within the block reported that Lonsdale had returned home at about 3.00 p.m. At 3.45 p.m. Lonsdale donned a raincoat but no hat and Elwell watched the suspect drive off in his Studebaker towards Waterloo.

Earlier that afternoon, a man in a cloth cap had joined the knot of travellers standing before the enormous indicator board at Waterloo station. As he gazed up at the arrivals information there was an echoing announcement from the loudspeaker. The 12.32 train from Salisbury was delayed and would be at least thirty minutes late. The man in the cloth cap walked to another in a bowler hat leaning on an umbrella and standing near platform 14, where the Salisbury train was expected to arrive, to pass on the information. He in turn walked to a 'newspaper seller' at the exit to Waterloo Road. Finally, the information made its way to a small car park on a patch of bombed-out land opposite the Old Vic theatre. Here dealers sold second-hand cars, and in the front line was a 1957 green Morris Oxford with 'For sale' white-washed on the windscreen. Inside was the jovial, bespectacled figure of Superintendent George Smith of Special Branch and an MI5 officer, pretending to be potential customers, while outside another man was inspecting the bodywork. The car was in fact in radio contact with the three other Security Service cars parked and waiting in neighbouring backstreets to make the planned arrests. Also scattered at strategic points around Waterloo station were fifteen A4 watchers to ensure that the police officers were brought to the rendezvous point at the crucial moment, immediately after (MI5 hoped) Houghton had handed sensitive naval secrets to Lonsdale. Many sported disguises: some were dressed as porters, while Superintendent Smith himself wore a black French beret.

The Salisbury train hissed to a halt at platform 14 at 3.15 p.m. and Houghton and Gee stepped down from their carriage, Gee clutching a straw shopping bag. Around the platform the watchers in their various disguises stiffened to alertness. The couple were tailed out of the station's main exit, with Gee glancing over Houghton's shoulder from time to time to check if they were being followed. Unexpectedly, they halted just as a number 68 bus was about to pull away from a stop and stepped onto the open running board at the back. There was subdued panic among the surveillance team. Only one Security Service watcher was close enough to take action. He sprinted to match the gathering speed of the bus and sprang on board. He overheard Houghton tell the conductor that he wished to visit East Street market, and told another passenger he wanted to do a little shopping there before returning to Waterloo and catching a train back to the country. The journey to the market lasted fifteen minutes through the traffic-clogged streets. In 1961, on a Saturday afternoon the East Street market in Lambeth was a noisy and bustling thoroughfare of stalls with hawkers shouting to advertise their wares. The MI5 watcher followed Houghton and his girlfriend more closely than he would have wished and with a sense of helplessness:

he could not communicate with his colleagues back at Waterloo station, who he knew could not move from their set positions at the station to assist him if Houghton and Gee disappeared from view into the crowd. The couple wandered aimlessly among the stalls. The watcher sighed inwardly with relief when Houghton and Gee ambled back to Walworth Road and caught another bus back to Waterloo, where they entered a small Italian café at 4.15 p.m.

At 4.25 p.m., the watchers discovered Lonsdale's Studebaker parked in a backstreet and Lonsdale himself was spotted briefly near the Old Vic studying posters for that night's performance of *A Midsummer Night's Dream*. Houghton and Gee ambled out of the café at the same moment and crossed over towards the Old Vic. On the opposite pavement stood Superintendent Smith, now joined by Ferguson Smith, back from Salisbury, and Special Branch officers. The watchers gave a prearranged signal to the Special Branch officers to identify Lonsdale, who was now waiting on the pavement outside the Old Vic. The policemen fanned out. Houghton and Gee approached Lonsdale, exchanged glances, and Lonsdale fell in a few paces behind them as they walked south towards St George's Circus. Superintendent Smith followed. After a short distance Lonsdale caught up with the couple, stepped between them, threw his arms around their shoulders as they exchanged friendly greetings, and took Gee's shopping bag.

At this instant, the superintendent jumped in front of them, shouting out: 'You're under arrest. I'm a police officer.' Gee and Houghton froze, Gee merely gasping out 'Oh' and Houghton 'What?' Lonsdale was silent. The Special Branch and MI5 officers lunged forward to grasp the prisoners by their wrists in what the Security Service described later as 'a very sure and firm grip' to prevent them destroying evidence, and to separate them from each other by six to eight yards. At the same moment three MI5 cars swerved to a halt beside them. Recognising that Lonsdale was the most valuable prisoner, Smith snatched Gee's shopping bag from his grasp and bundled him into the nearest car. As soon as they were speeding towards the Thames Embankment the superintendent pulled the radio microphone to his lips and, with a barely suppressed smile of self-satisfaction, announced in his Wiltshire burr, 'Lock, stock and barrel' – the code sign to MI5 and Special Branch that the three suspected KGB agents had been successfully arrested.

Within seconds, Winterborn and two Special Branch officers strode across from the A4 observation post, which overlooked The White House, to Lonsdale's flat and searched it quickly. They focused on the hiding places the Security Service were already aware of: the hidden compartment inside Lonsdale's cigarette lighter (where they found the same incoming and outgoing OTPs minus a few sheets), and the Chinese scroll hanging above the bed (on opening the secret cavity in the bottom roller they discovered three rolls of bank notes totalling 1,800 US dollars). Elwell and Winterborn then rushed over to Special Branch on the Thames Embankment.[90]

Back at New Scotland Yard, Smith checked Gee's shopping bag and was relieved to find four pamphlets from the Portland base giving confidential details of research tests on an underwater sonar, and a sealed tin containing undeveloped camera film, which he immediately handed to MI5 to process. Gee, Houghton and Lonsdale had been caught red-handed in an act of espionage.

The three prisoners were placed in separate rooms to be questioned. Smith told Lonsdale – who was instructed to strip down to his underwear – that he was being held on suspicion of having committed offences under the Official Secrets Act, cautioned him, and asked his name and address. In response he said, 'To any question that you might ask me my answer is no. So you need not trouble to ask.' Special Branch found in Lonsdale's pockets £205 in £5 notes and fifteen twenty-US-dollar bills – almost certainly money to be given to Houghton and Gee. He went on to make a request: 'As I appear to have to stay here all night, can you find me a good chess player?' Such was the prisoner's charm that the superintendent changed the spy's escort to include a policeman with knowledge of the game.

There was only time to question Lonsdale, Houghton and Gee briefly. Houghton blurted out that he had been 'a bloody fool'. Ethel Gee expressed surprise at her arrest, murmuring 'I have done nothing wrong.'[91]

At around 6 p.m., Superintendent Smith, Chief Inspector Ferguson Smith and other Special Branch officers climbed into three cars and drove out west through the wintry gloom of Notting Hill, Shepherd's Bush and the A40 road towards Ruislip. Following behind in their car were Elwell and Winterborn. Elwell's plan demanded they arrive at 45 Cranley Drive around the time the Krogers would have expected Gordon Lonsdale on the Saturday night with his latest batch of secrets.[92] They all arrived shortly after 6.30 p.m. (the police three minutes after MI5 because they lost their way in the network of streets around the Krogers' house), parked in the cul-de-sac where Lonsdale usually left his car, and doused their lights. Extra police officers were waiting, discreetly hidden in delivery vans or cars. At a signal from Smith they quietly fanned out round the property to prevent any escape through the back door and garden.

All was silent and quiet for a few moments under the sparse streetlights. The suburban road was bare and deserted. During the day the Krogers had been observed closely from the 'Dust Cover' surveillance post for any sign that they might be suspicious. Helen Kroger went out shopping in the morning in what the Searches described as her 'town clothes', as opposed to her usual slacks and sweater. At 4.50 p.m. there had been a brief bout of whispering and suppressed panic when she crossed the road to 1 Courtfield Gardens, knocked on the door and spoke with Ruth Search for a few minutes before calling on another neighbour in Cranley Drive, where she stayed for about half an hour before returning home.

Superintendent Smith was understandably tense as he stood in the sheltered alcove at the front entrance of 45 Cranley Drive with Ferguson Smith and rang the doorbell, fingering nervously the search warrant in his pocket. If the Krogers were suspicious they might refuse to let him in and immediately start to destroy any evidence. Smith knew that the doors and windows were all secured with robust locks: it would be difficult to smash their way in and would attract extremely unwelcome attention. He waited patiently for a few seconds. Through the coloured glass panel in the door he saw the hall light flick on and someone approach. A bolt rattled, a lock turned, and the door was opened. Peter Kroger stood in the entrance smiling, pleasant, and slightly puzzled. Smith touched his hat and said he was a police officer making enquiries about some recent burglaries in the area and could he come in for a brief chat. Kroger waved them into the sitting room and a few seconds later Helen Kroger appeared. The superintendent then ignited his conversational explosive.

He introduced himself as Superintendent Smith of the Metropolitan Police, and asked them to tell him the name and address of the man who came and stayed at weekends, particularly the first Saturday every month, and arrived about 7.15 p.m. After a few seconds of stunned silence, Helen Kroger recited a few names. Lonsdale was not among them. Armed with this evidence of lying about their associates, Smith announced he was arresting them and asked the Krogers to get ready to leave for the police station. Helen Kroger strode out of the sitting room and into her bedroom. A woman police sergeant followed and saw her take out a coat from a cupboard, which she put on, and then snatch up a brown hand-bag from a chair. Helen Kroger casually told the superintendent she just wanted to stoke the boiler in the kitchen before leaving and started to move towards the door. Smith blocked her exit, and insisted on looking in her handbag first. The superintendent tried to take it from her but Helen Kroger gripped the bag tightly, and the catch sprang open. Smith seized it and within he discovered a white envelope containing a typed sheet of cipher, a six-page let-ter written in manuscript in Russian, two glass slides with three microdots sandwiched between them and the list of eight roads first discovered in Lonsdale's cigarette lighter in September. Helen Kroger immediately declared in her husky North American accent, 'I know nothing about this.' Police officers moved to start a search of the loft while the two suspected spies waited to be driven to a local police station. Peter Kroger was remarkably cooperative. He found for the officers the stick to pull down the loft ladder, and even clambered up the ladder himself in the dark to grope for the light switch. Further cooperation ceased almost immediately, however, at Hayes police station.[93]

The Sunday after the arrests Security Service intercepts of phone calls recorded the mounting concerns and dismay of Lonsdale's busi-ness associates. A solicitor friend went to New Scotland Yard that morning to have 'a long chat' with the suspect. However calm on the outside, Lonsdale was clearly in turmoil within. The confused and rambling story he told the lawyer about the meeting with Houghton and Gee being one of chance did not impress him. 'I don't believe what Gordon told me this morning for one moment,' the lawyer confided to Lonsdale's business partner, Michael Bowers, over the telephone, 'what he told me was absolute balls, quite honestly.'

While Special Branch and MI5 started to quarter the Krogers' bungalow for further evidence and clues, Superintendent Smith, never timid of publicity and much to the chagrin of the Security Service, was already confirming the story to his media contacts. The arrests of the alleged Soviet spy ring led the early-evening BBC television and radio news bulletins on the Sunday night. Those who thought they knew the Krogers well and socialised with them were astounded.[94]

The contents of Helen Kroger's handbag were swiftly analysed. The three microdots proved to be letters to Lonsdale in Russian from his family: three from his wife dated November and December; one from his mother written on Christmas Day; and one from his daughter. The manuscripts in Russian were Lonsdale's replies to his wife and children. Elwell reconstructed events on the Saturday of the arrests. Lonsdale had received the three microdots on the Saturday morning (reading them by microscope), written by hand his responses to his wife and children, and typed out a coded message to be sent to Moscow. He handed these three items to Helen Kroger in his leather briefcase at a clandestine meeting in the middle of the day, probably at one of the locations marked on the list of eight roads first discovered in Lonsdale's cigarette lighter in September and found in her handbag when arrested. Elwell speculated that Lonsdale needed to pass the items to Helen Kroger as a precaution before his meeting with Houghton and Gee at 4.30 p.m., and to allow time for the Krogers to prepare Lonsdale's responses to his family letters ready for him to take to a dead drop on the Sunday.[95]

Like Lonsdale, the Krogers initially refused to allow the police to take their fingerprints. Frustrated, the police and the Security Service waited until Lonsdale and the Krogers appeared at Bow Street magistrates' court two days after their arrest and applied for orders for them to be fingerprinted. Lonsdale relented but the Krogers continued to object violently, Helen Kroger asserting, 'We are not criminals, so why should we give our fingerprints?' The magistrate, however, granted the order, and within minutes, in the bleak prisoners' waiting room at the court, a New Scotland Yard police officer was pressing their fingers onto an inked pad and then onto a printed form. A detective was soon back at the Yard's Criminal Record Office comparing the Krogers' prints with those already stored there. Within a day or so, five sets of prints were isolated with similar characteristics. Three were eliminated, leaving just two sets which matched perfectly. They were from a file marked 'Espionage suspects' sent to New Scotland Yard in February 1958 by the FBI. The prints belonged to a couple named Morris and Lona Cohen. The FBI had been hunting around the world for these suspected Soviet spies for the past three years.[96]

Part Two
Morris and Lona Cohen

5

The FBI hunt for the Cohens

Although the FBI had been on the trail of the Cohens since 1957, its hunt for the couple had in fact first begun in New York City four years earlier, in 1953. In October that year an 'informant of unknown reliability' but who was 'in a position to know' had told the Bureau that Morris Cohen and his wife were 'very radical communists' and had held Communist Party meetings in their home. Classified as a 'Security Matter C' (meaning communist) and marked 'Secret', the report summarised the results of five days of preliminary enquiries. FBI agents discovered that the Cohens had disappeared from their New York address two or three years before without leaving any forwarding address. They were not on the files of the New York police. The FBI had no photographs of the couple. The New York division sent the report to FBI headquarters in Washington in November 1953, and its espionage squad continued to investigate the Cohens.[1]

In 1953 the Director of the FBI, J. Edgar Hoover, was in his anti-communist pomp. Since boyhood in Washington, this intense and competitive personality had devoted himself to protecting the order and values that middle-class, white, Protestant America cherished against the forces of change and turbulence. The chief threat to that order and those values in Hoover's eyes was communism. He viewed it, in the words of one distinguished historian of the FBI, as 'the sin as well as the crime of the century'. A month after that first report about the Cohens arrived at the Bureau's headquarters in Washington, its director was presenting the various Soviet spy cases the FBI was pursuing as part of a Manichean battle between those like the FBI who defended traditional American values, and those who sought to undermine them: 'Our American way of life, which has flourished under our Republic and has nurtured the blessings of a democracy, has been brought into conflict with the godless forces of communism.

These Red Fascists distort, conceal, misrepresent and lie to gain their point. Deceit is their very essence . . . to a Communist there are no morals except those which further the world revolution directed by Moscow.'[2]

One of Hoover's chief lieutenants in the Bureau's war against communism was Alan H. Belmont, head of the FBI's Domestic Intelligence Division since the end of 1951. His job was to oversee the FBI's work in keeping America safe from fascists, subversives, communists and Soviet spies, and he was to supervise the FBI's hunt for the Cohens in the years ahead. Born in 1907 in the Bronx, he was three years older than Morris Cohen and six years older than Lona. From a family of modest means, he had worked his way through college in California and was awarded a degree in accounting from Stanford in 1931. During the years of the Depression that followed he took a variety of temporary jobs until recruited by the FBI in 1936. He started investigating theft and extortion cases in Alabama and in his early days in the Bureau was described as 'somewhat timid'. Belmont soon overcame his natural reticence and was moved to senior positions in FBI offices in Chicago and Cincinnati. Belmont was, as a wartime report on his performance noted, intelligent and 'a product of hard work'. In 1944 he was promoted to be assistant agent in charge of all security work in New York. It was in this role that he learnt about Soviet espionage in America and worked successfully on several high-profile investigations and prosecutions.[3]

In February 1946, Belmont was moved to take charge of New York FBI Division I, which handled 'all security matters, including communist cases and Russian espionage'. In light of the increasingly menacing international situation, this was some of the most important work being handled by the Bureau. Russian espionage was only one of six sections within the division and was overwhelmed with work, having 217 open cases in September of that year. Before Belmont started his new job, the FBI counter-intelligence team in New York (in particular a young and dynamic agent called Robert Lamphere) collected tantalising scraps of information about Soviet espionage in America, but nothing coherent or solid in terms of evidence. This changed with two crucial defections. Igor Gouzenko, a Soviet army intelligence cipher clerk who defected in Canada in September 1945, provided the first proof that the Soviets had penetrated the Manhattan Project to develop the atom bomb, and controlled highly placed agents in the US government. A little later an American woman spying for the USSR called Elizabeth Bentley walked into the FBI's New Haven offices and provided the Bureau with the names of more than eighty agents. At last the FBI possessed a detailed knowledge of Russian espionage rings in Washington and New York. Unfortunately, when Hoover relayed Bentley's information to the MI6 station chief in Washington, he passed it to his deputy. That deputy was Kim Philby. He in turn fed this invaluable intelligence to the Russians, who immediately began to roll up their compromised spy networks.[4] In early 1948 US army code-breakers started to provide the FBI with fresh insights into Soviet espionage which were to prove even more valuable than those passed on by Gouzenko and Bentley. Over a period of months and years they decrypted top-secret messages sent during wartime between the US consulate in New York and KGB headquarters in Moscow. It was known as the Venona Project. Astoundingly, largely at the insistence of Hoover, the US army shared this highly secret intelligence only with the FBI and the British, not with President Truman or, until 1952, the fledgling CIA.[5]

Only someone with Hoover's longevity could have placed the domestic consequences of the Cold War in a historical perspective. With a personal memory of the paranoia generated in the United States in the early 1920s by the Bolshevik Revolution, Hoover pushed successfully for the prosecution of leading members of the Communist Party of the USA (CPUSA) under the Smith Act (which banned anyone from organising or joining any organisation aimed at the violent overthrow of the government). Belmont helped arrest and prosecute the eleven CPUSA leaders who were put on trial in November 1948 in the Foley Square courthouse in New York, near the old offices of the FBI. After one of the most contentious trials in American history, the jury found all twelve guilty eleven months later. The trial laid waste to the CPUSA, which was emasculated as a political force and an organisation capable of providing succour to Soviet intelligence operations in America.[6]

Belmont, meanwhile, was leading the hunt for Soviet spies in New York uncovered by the decoded Venona messages. Two of the most important were Alger Hiss and Judith Coplon. Coplon worked in the Justice Department. On a trip from Washington to New York in March 1949, Belmont supervised the tense surveillance of Coplon when she met her KGB controller and passed him a document which she thought was highly sensitive but in fact had been sent to her deliberately as bait by the FBI. The couple were arrested. Belmont gave evidence at Coplon's two trials; she was found guilty on both occasions and Hoover commended Belmont on his work. The convictions, however, were overturned on appeal because of investigative illegalities involving wiretaps and arrests without warrant. Hoover was outraged and searched for scapegoats. His anger was fuelled by developments in another, and ultimately much more significant, spy hunt which was also to entangle Belmont.[7]

In mid-September 1949 a deciphered Venona message revealed to a startled FBI that in 1944 the KGB had an agent within the British mission to the Manhattan Project. When, a few days later, President Truman announced to a troubled world that the Russians had exploded an atomic device, the FBI had already isolated a prime suspect: Klaus Fuchs. Then back in England, Fuchs finally confessed to MI5 in December 1949. He disclosed an intriguing fact as part of his confession. A man known to Fuchs only as 'Raymond' was the conduit he used for passing information to the Russians. This revelation coincided with President Truman proclaiming that, because the Soviets now possessed the atomic bomb, the USA would build the hydrogen bomb, and Senator Joseph McCarthy making his speech in early 1950 claiming to possess a list of 205 communists in the US State Department. In this electric political atmosphere, Hoover demanded with his customary fury that the FBI find 'Raymond' at all costs. Most of the Bureau's field offices and agents were mobilised. One high-profile victim – unjustly – of Hoover's anger was the Assistant Director of Domestic Intelligence, who was dismissed. Belmont had taken a day off in February 1950 and brought his newly adopted daughter to work to introduce her to colleagues when his telephone rang. It was Hoover, asking him to come to Washington as an FBI inspector to oversee the Domestic Intelligence Division. Belmont knew that Hoover was not content with its operations, and that the work of the division was controversial, attacked both by the liberal left in America for hounding communists real and imagined, and by those who did not believe the intelligence-gathering activities of the Soviet Union and its satellites posed any substantive threat to the USA. Belmont coveted neither the job nor a move to Washington, but, as he wrote candidly later, 'what do you say when the man in charge of your outfit wants you to do a job? I told Mr Hoover, "As long as I am working for you, I will take on any assignment you feel is necessary."' Belmont knew Hoover was not a man to be crossed.[8]

Belmont, predictably in this volatile atmosphere, was regularly berated by Hoover. Belmont recalled later, 'For months after arrival in Washington I practically lived at the office, and it was not possible until several months later that I was able to bring my family down from New York to take an apartment.' Despite his ceaseless efforts to sharpen the FBI's security work, he was criticised harshly by the internal FBI inspectors unleashed by Hoover, who Belmont was convinced 'were under instructions to get' him. Incensed, he considered resignation. But the wind changed abruptly in early 1951. Criticisms mollified and faded. Belmont remained in the Bureau, his loyalty to Hoover cemented rather than eroded by the harsh treatment. The pattern of his life at the FBI was set: long stretches of exceptional diligence interspersed with occasional censures for 'delinquency' by Hoover (for example, for failing to pass on information to the British about the KGB spy, Bruno Pontecorvo, who defected to Russia in September 1950), which in turn Belmont helped overcome by judicious flattery of the Director. The texture of his personal life was also calmer: Belmont and his spouse had settled in Washington and now had a daughter and a son. He was promoted to Assistant Director of the Domestic Intelligence Division in October 1951.[9]

Undoubtedly key to Belmont surviving in Washington was the discovery of 'Raymond'. Clues from decrypted Venona messages and detailed work in the files pointed to one suspect, a chemist called Harry Gold who lived in Philadelphia. Finally, at the end of May 1950, having been shown new moving pictures and photos of Gold, Fuchs identified him as 'Raymond'. Gold confessed and, under Belmont's supervision, the FBI was soon running forty-nine investigations based on his revelations. One was into a man whom Gold thought was his contact at the Manhattan Project, David Greenglass. Greenglass in turn confessed after his arrest and said he had been recruited as a Soviet spy by his sister

Ethel's husband, Julius Rosenberg. At last the FBI understood that the individual engineers at Los Alamos suspected of spying for the Soviets were in fact planets revolving around the sun of the espionage network, Julius Rosenberg. Julius and Ethel Rosenberg were both sentenced to death. Many in the FBI expected that Julius would cooperate with the Bureau to save his wife. Both remained silent, as did Moscow. On the day appointed for their execution, 19 June 1953, it was Alan Belmont who secretly travelled out to the death chamber at the Sing Sing Correctional Facility in New York in case the Rosenbergs decided at the very last moment to save themselves by confessing. As the time for the executions, 8 p.m., ticked closer, Belmont sat by an open telephone line back to Washington. A few minutes before the appointed hour, he telephoned to say that the Rosenbergs had for the last time refused to confess. It was a day Alan Belmont was never to forget. He knew the Rosenbergs were guilty of espionage but, like other FBI colleagues who had worked on the investigation, bore a sobering responsibility for their deaths.[10]

When that first report on the Cohens arrived in Washington in October 1953 it would not have attracted Belmont's attention immediately. There were other intelligence matters of higher priority: for example, Senator William Jenner was investigating the Soviet informant Harry Dexter White, and Senator Joseph McCarthy was about to examine alleged subversion at the Signal Corps research centre at Fort Monmouth in New Jersey (hearings which were to lead to McCarthy's downfall in 1954). So back in New York, the espionage squad plodded on with their investigation into the Cohens as winter set in. The agents in the field usually worked in pairs, smartly dressed in the informal Bureau uniform of white shirt, tie, business suit and snap-brim fedora, with a .38 pistol in their holster. Each agent took notes, and when preparing their report compared the information they had gathered for accuracy. In the 1950s all agents were male (Hoover, as Director, refused to allow women to become agents – although behind his back, pragmatic subordinates like Belmont often used women described as 'auxiliaries' in espionage cases), and only a handful of agents in major field offices conducting investigations were black. FBI agents were imbued with a sense of pride at belonging to an elite institution but also with fear: fear of failure, and the ever-present fear of incurring the displeasure of Hoover, whose centralised reign over the Bureau was a form of 'one-man rule'.[11]

The New York office made little progress until a source reported that Morris Cohen had served in the army during the war. In February 1954, agents visited the Veterans Administration office at 252 Seventh Avenue. There the librarian dug out Morris Cohen's army file; at last his ghostly figure began to take on form and substance. The file provided the espionage squad with a summary biography together with other information which the FBI investigated in the years that followed. Born in July 1910 in East Harlem, Morris Cohen was the son of Harry and Sarah, recent Jewish immigrants from Ukraine and Lithuania. Harry had started peddling fruit on the streets, before establishing a shop and moving his family north-east to the rapidly expanding and more verdant suburb of the Bronx. Educated in the Bronx, Morris Cohen had received a degree in English from Mississippi State College in 1935, fought in the Spanish Civil War in the communist-influenced Abraham Lincoln Brigade, and on his return to New York worked for a news company and a cafeteria between 1936 and 1942 but also (intriguingly, according to an FBI report) 'for the Russians during the New York World Fair during 1939 and 1940'. The fair, at Flushing Bay, covered a site of nearly two square miles. One of the most prominent pavilions was that of the Soviet Union, featuring massive stone medallions of Stalin and Lenin and a 'tall marble and porphyry pylon, topped by a stainless steel statue of a Russian youth holding aloft a five-pointed star'.[12] There could have been no clearer indication of Cohen's political sympathies. In July 1941, in Norwich, Connecticut, he married Lona Petka. She was three years younger than Morris, born in January 1913 and brought up in Connecticut farm country by Polish immigrants.

Cohen joined the US army in July 1942 as a cook (conscripts who had fought with the Abraham Lincoln Brigade were sometimes regarded with suspicion and given lowly roles because of security fears), served in Europe after the 1944 Normandy landings and was discharged in November 1945. Afterwards he trained as a teacher at Columbia University in New York, and taught at Public School 86 for a year starting in September 1949. The final address for the Cohens in New York was Apartment 3B, 178 East 71st Street, which they had rented from November 1943. Their Upper East Side home was in a tired but respectable brownstone building, demolished in the early 1960s, four blocks from the elegant blue-bloods and solid burghers of Central Park to the west.[13] It was from here that Morris Cohen sent a letter to his school on 29 August 1950 explaining that he was resigning his post suddenly because he had obtained 'a very desirable writing job'. One source suggested that the Cohens had moved to California. Around this time the FBI also obtained photos of the couple for their file. The espionage squad agents returned to their new offices in the former Dun & Bradstreet building at 290 Broadway in lower Manhattan, a brisk walk from Brooklyn Bridge, to prepare their reports for Washington headquarters. With its numbers swelling from 700 to nearly 1000 agents in the face of the burgeoning number of security and criminal cases, FBI New York division had recently outgrown their cramped accommodation in Foley Square.[14]

In October 1950, Belmont assembled for the first time a lengthy memo for Hoover summarising all the information about KGB spies operating in New York in 1944–5 as it had emerged from the decrypted Venona messages. Belmont's team revised and reissued the memo every six months in the years that followed, collating the names of the spies who were prosecuted, cooperated or remained simply code-names, tantalisingly unidentified. This last group, unfortunately for Belmont, included the Cohens. As the year of the Rosenbergs' execution ended and 1954 began, FBI field offices across the USA followed up on the various leads on Morris and Lona Cohen generated by the Veterans Administration file. In New York, on a pretext, an agent contacted Morris Cohen's father, Harry, and later Cohen's mother, but both said they did not know of their son's whereabouts. Nor did Harry's brother, Abner, who lived in New Jersey. Enquiries were made in California and in Lona Cohen's home area in Connecticut, but to no avail. All the FBI garnered were scraps of rumour and information of disparate value and relevance, for example that the health of Morris Cohen's mother was declining (she died in March 1955), and that Cohen had been receiving a very modest monthly disability cheque from the government. Many US Communist Party informants in New York professed ignorance of Morris Cohen. It is clear from the Cohens' FBI file that Alan Belmont monitored the investigation closely, receiving and initialling intelligence reports when he received them in Washington. However,

there was no particular reason to prioritise the case, which meandered forward.[15]

Further details accumulated like sediment. In May 1955 two FBI
agents interviewed Harry Cohen at his apartment in the Bronx. He
told them that his son and his wife had left New York and gone to
California about five years before. In the first few weeks that followed
he and his wife received some letters, but he had heard nothing from
him since. Harry Cohen said his son's long and unexplained absence
had been a cause of considerable concern and probably contributed
to his wife's recent death. New York continued to coordinate enquiries across the United States throughout 1955 and into 1956. By
October that year it was clear the investigation had lost impetus and
New York informed Washington that they considered the Cohen
case closed, subject to any further information coming to light.[16]

During the 1950s the FBI headquarters in Washington was in a
forbidding granite building in Constitution Avenue shared with the
rest of the Justice Department. Within, as with MI5's Leconfield
House, the offices were cramped. Many of the more spacious rooms
were divided by glass or steel partitions occupied by as many as four
agents and four secretaries. Belmont, as an assistant director, of
course had his own room. He aimed to get involved with and lead
those working for him. 'My door was always open and employees
were encouraged to bring problems to me,' Belmont wrote later.
Although he had a number of detractors inside and outside the FBI,
Belmont was widely respected by agents and supervisors within the
Bureau for being hard-working, tough, intelligent and pragmatic:
his pragmatism showing itself in a willingness to be flexible when
enforcing the draconian rules and regulations of the Bureau. He also
had a self-deprecating sense of humour, illustrated by the story he
told of a White Slave Traffic Act raid on a brothel near Chicago in
1937. One prostitute tried to escape through an open window but
Belmont managed to seize her by the leg and drag her back. Furious,
she turned on him and spat out, 'You G-men make a lot of fuss but
you'll never make sex unpopular.'[17]

In 1957 crucial further information surfaced unexpectedly which
caused Belmont to jolt the comatose Morris and Lona Cohen investigation back into vigorous life. Soon after dawn on 21 June in
New York, FBI and US immigration officers arrested a man suspected of being a Soviet spy at the Hotel Latham at 4 East 28th

Street under the name of Emil Goldfus. He soon told the FBI he was in fact Rudolf Abel, and it was under this name that he was to be prosecuted as a Soviet spy. In Abel's hotel room a package was found containing 4,000 dollars and, under a flap of the paper in which the money was wrapped, two photographs of a man and a woman. Written on the back of the woman's photograph was 'How are Joann's murders?', signed 'Shirley', and on the back of the photo of the man, 'Have you taken a trip to Frisco since we last saw you?' and the signature 'Morris'. When interviewed, Abel mumbled that the photos were of a couple he had met in Central Park in 1949. He knew them only as Shirley and Morris, and had associated with them for about four months, meeting them around twelve times during this period. Asked if he had ever used them in his espionage work, Abel replied, 'No'. Understandably, Belmont and his agents were unconvinced. The photographs were quickly identified as Morris and Lona Cohen. The Bureau speculated that the questions written on the back might be 'paroles' for a meeting between Abel or one of his agents and the Cohens, and the Cohens might be linked to the Rosenbergs.[18]

The louring figure of the FBI Director, J. Edgar Hoover, ordered the Bureau's principal field offices to treat the case as a priority: all leads were to be followed up within five days and Washington informed of important developments 'by the most expedite means'. Belmont, sensing the heat of Hoover's breath on his neck, organised a renewed and now urgent manhunt for the Cohens across America, from California to Connecticut. For the first year Belmont put in charge two agents who had a reputation as 'a couple of real tigers'. Lona's sister, Sophia, told the investigators that the Cohens had held a farewell party at their apartment at the start of July 1950 to celebrate their impending trip to California, where Morris was to work in a private school. They planned to leave their belongings and send for them if they decided to remain on the West Coast. She heard nothing more from them. Morris's father, Harry, contacted her in September to say they were not returning and together they visited the apartment. Sophia said: 'it appeared that Morris and Lona had taken just enough clothes for a weekend trip . . . Lona's jewelry, cosmetics, and personal effects appeared to be intact in the apartment. She said that Harry Cohen took part of the clothing and furniture and she took the rest. She . . . expressed uneasiness . . .

about taking the property, but he said that Morris and Lona would not be coming back.'[19] Morris had paid the rent for the apartment three months in advance.[20]

Descriptions and photographs were circulated. On reviewing the files, someone spotted that the Cohens disappeared suddenly from their New York apartment in 1950 a few months after the FBI made their first arrests of KGB agents linked to the network of Soviet spies controlled by the Rosenbergs. Lona had closed her savings account on 14 June 1950, and Morris Cohen two days later – the same day David Greenglass (brother of Ethel Rosenberg) was arrested. The concern was that, after fleeing New York, the Cohens 'entered the CP [Communist Party] underground . . . [and] have been involved as espionage agents for Russia since that time'.[21] The photographs of Morris and Lona Cohen were shown to the convicted American atom spies, Harry Gold and David Greenglass, in jail, and to other associates of the Rosenbergs, but they professed ignorance. Communist informants in New York were questioned and background details about the Cohens dutifully assembled and sent to Washington. Some reports were not much more than gossip: 'Lona Cohen definitely believed in free love and had very loose morals,' and 'incessantly dressed in mannish attire' by wearing 'slacks the year round' (coincidentally, many of Lona Cohen's neighbours in Ruislip were remarking on the same sartorial singularity).[22]

Others helped to develop the biographies of the Cohens. Lona was the daughter of cotton weavers and had been brought up in the mill town of Adams, Massachusetts. In an early sign of her headstrong nature, she had left home and school in Connecticut aged fourteen and came to New York on her own, working first as an underage factory hand in Manhattan. From an early age she was drawn to radical politics. 'She was always talking about justice and inequality,' a younger sister later recalled, 'she wanted to change the world.' The FBI learnt that she moved to an apartment in Greenwich Village where, according to an informant, she had a 'hell of a good time'. At this time Lona Cohen sampled various brands of left-wing politics before she joined the Communist Party in 1935, aged twenty-two. In 1936 she was recruited as a governess to look after the young son and daughter of the wealthy Joseph Winston in Park Avenue, Manhattan, where she worked until her marriage in 1941.

Lona developed an especially close relationship with the Winstons' son, Alan, and there were suggestions that she tried to proselytise by giving him copies of a communist newspaper. She stayed in contact and was later fondly remembered by him as a 'tomboy' who enjoyed throwing a baseball and skating in Central Park.[23]

Just over a week after the arrest of Abel, a confidential source revealed to the Bureau that Alan Winston had in April 1957 rented a safety deposit box for a year at the Manhattan Trust Company in Third Avenue in New York. The next day at 10 a.m. the box was opened under a search warrant and inside were four packages wrapped in brown paper containing a total of 15,000 dollars in cash. On each parcel was written the name 'Milton'. The same day Winston, then aged twenty-seven and a student teacher, was interviewed and shown photographs of Rudolf Abel, whom he identified as someone he knew as 'Milton' or simply 'Milt'. Later Winston (who denied being a communist himself) identified the photographs of 'Shirley' and 'Morris' as the Cohens and told a tangled tale to the FBI which conclusively linked Abel to the Cohens.

Winston had visited the Cohens in their apartment in the late 1940s, but one day, when he arrived in June 1950, Lona told him that she and Morris were going away for a short time and gave no reasons or destination. In the apartment were fifteen or twenty boxes crammed with their possessions, and Lona asked him to take care of their storage while they were absent. Winston agreed, and that was the last time he saw the Cohens. Later that year he received a letter from Lona asking him to meet her at a spot near the Central Park Zoo, which Winston knew from childhood visits with her when she had been his governess. There he was met by 'an elderly man' who introduced himself as 'Milt' and explained that Lona could not make the meeting and had asked him to come in her place. The two struck up a friendship based on a mutual interest in art and met every month until Winston joined the army in 1953. From 1950 onwards, Winston continued to pay the monthly storage charges for the Cohens' possessions. Whenever he was running short of money, 'Milt' stepped in to assist. Around the end of 1956 Winston said that 'Milt' had hand-delivered a second letter to him from Lona Cohen without an envelope 'written on thin onionskin paper' which contained nothing but vague pleasantries; he never answered this letter and did not ask 'Milt' any questions about it. His attitude throughout appeared to the FBI strangely naïve and uncurious. Although Winston admitted with hindsight 'that some of his activities were foolhardy but that at the time he believed he was merely doing favors for a friend', the FBI were clearly unconvinced, not least because they were unable to locate the depository where the Cohens' belongings were allegedly stored.[24]

After this somewhat grand life near Central Park working for the Winston family for five years, Lona had plunged back into an existence shared by numerous New Yorkers at this time: she worked in an arms factory. Hers was on Long Island and manufactured parts for aircraft. She suffered an industrial injury to a finger, was a union leader and according to a manager at the factory 'followed the communist line'. Sources disclosed to the Bureau that Lona transferred from the industrial to the Bronx section of the Communist Party in New York in October 1943, and that she had 'dental work done on her front top teeth and the job had not been a good one', leaving her with characteristic buck teeth.[25]

Morris Cohen clearly had communist leanings, imbibed from his parents. One family member, using Morris's Yiddish name, later declared that 'Moishe got communism as his mother's milk'. More details about his life during the 1930s Depression emerged: how back in New York City he joined the Communist Party in late 1935, almost certainly operating under a false name, and quickly became a branch organiser in the Bronx. The FBI garnered further information about Cohen's pilgrimage to war-torn Spain. It emerged that back in 1937, when he had joined the Abraham Lincoln Brigade to fight in Spain, with the help of communist International Brigade recruiters he had taken on the identity of another potential but rejected recruit. Although Cohen already had a passport of his own, in 1937 he was issued with a second one in the name of Israel Altman, with a photograph of Cohen 'in a crude attempt to disguise himself' wearing horn-rimmed spectacles and a moustache. The FBI learnt that while Morris Cohen worked at the New York World Fair in 1939–40 a Russian called Mikhail Svirin was employed there – later to be Abel's KGB contact in the Russian embassy. The investigators were not spared even the most intimate details of his life, including that while serving in the army in 1944 Morris underwent a lengthy treatment for haemorrhoids, a condition that was to trouble him for the remainder of his life.[26]

One friend of Morris Cohen from his wartime service, a doctor, painted an insightful picture of the Cohens as a couple. He visited them frequently at their East 71st Street apartment after the war. He characterised Morris as

> an extreme realist . . . a highly intelligent individual but . . . an introvert . . . He stated that Lona Cohen was an extreme extrovert and was therefore very good for Morris. He stated that Lona Cohen would usually take a dominant stand at the outset of any discussion but that she would be placed in a secondary position in the final analysis, as she did not have the intelligence to compete with arguments put forth by Morris Cohen [whose] only hobbies that he knew of were reading and boxing. He stated that Cohen read anything available, whether it was a book, pamphlet or newspaper.[27]

Morris Cohen's brother-in-law provided other intriguing evidence. He had met his wife, Lona's younger sister, through Morris. But he and Morris were not close, largely due to their different politics. He knew that Morris and Lona were 'fanatical' communists and, as a trades union representative himself, quarrelled violently with his brother-in-law in 1946 when Morris argued that the Communist Party 'should dictate the policies of the unions'.[28]

Despite the intensive manhunt, the whereabouts of the Cohens remained a mystery. It appeared that Morris sent letters to his father, Harry, but he never showed them to anyone. Lona's sister, Sophia, believed these letters in fact never existed but that the Cohens were in contact with their son through an intermediary.[29] The government disability cheques sent to Morris Cohen after his disappearance in the summer of 1950 were cashed until June 1954, almost certainly by his father, the signatures on all the cheques being identified by the FBI laboratory as matching Morris's handwriting. When Morris's mother, Sarah, died in 1955, relatives were puzzled and shocked that Morris did not attend her funeral. Harry swiftly remarried and died in December 1956. Again, Morris did not come to the funeral. The FBI believed that both parents had known far more about the disappearance of their son than they had ever told the Bureau; according to one source, although Harry Cohen did not know where Morris was living, 'an unknown man would appear at the door at infrequent and irregular intervals and give Harry Cohen money from Morris'. In 1956 the man came 'several times', and gave Harry 'substantial' amounts of cash and news that 'Morris was all right'. Although he was probably Abel, the mysterious visitor was never identified. The FBI speculated that around the middle of 1954, Morris Cohen 'had left his original hiding place, either in US or Canada and departed for Europe', where he may have 'met with foul play'.[30]

The moment had come for Belmont, puzzled and exasperated, to extend the hunt around the world. On 21 February 1958 the FBI sent out a note to US embassies in Bonn, Havana, London, Madrid, Mexico City, Ottawa, Paris, Rio de Janeiro, Rome and Tokyo to cover twenty-three countries. It attached fingerprints of the Cohens (on file in Washington since Morris joined the Army and Lona worked in an arms factory) and asked all US legal attachés to alert the police services in their countries to the search. The London attaché duly passed on the request to Special Branch at New Scotland Yard. Although the FBI knew the Cohens were contacts of Abel, they could not provide 'any precise idea of what function the Cohens performed'. The 1958 FBI request was just one of a great number which the British services received from various countries. Later that year, on behalf of the FBI, the Security Service requested checks of British records showing whether the Cohens had ever entered the UK. There was, unsurprisingly, a nil return other than them staying overnight in London en route to and from Paris in 1947.[31] When the fingerprints of the Cohens were sent to police around the world in February the reaction in the New York office was predictably cynical. One of the FBI agents involved, Robert Beatson, later recalled that there 'was a lot of good-natured joking . . . "Boy oh boy, that's like throwing a stone at the moon." I mean, what are the chances of anything happening from that? Slim and none.'[32]

Around 1958 Belmont put a new agent in charge of the hunt. By that stage, Beatson later remembered, the investigation was 'an old-dog case', namely a case in which there is very little hope of anything happening. The new lead agent and his team were determined to be thorough and diligently continued to pursue any leads. They checked all 158 former members of the Lincoln Battalion in the Lincoln-Washington International Brigade who returned from Spain to New York with Cohen on the steamship

Ausonia in December 1938 and former friends from his time in the army, and developed further the biographies of the Cohens as best they could.[33] Although the Bureau created numerous files crammed with information, the lives of the Cohens after September 1950 remained a stubborn blank.

The Bureau kept a close watch on associates of the Cohens in case they led them to their quarry. One was the veteran American communist Jack Bjoze, who had fought in Spain with the Lincoln-Washington International Brigade at the same time as Morris Cohen. At the end of the 1950s Bjoze was operating a travel business in New York with many communists and communist-front groups as clients. In December 1958, the FBI learnt that Bjoze and his partner were planning a trip to Bermuda early the next year. The Bureau asked British intelligence (which had excellent security contacts in the Caribbean) to 'cover their activities' while on the island 'to determine if any of their contacts might lead to information concerning the Cohens', explaining that they believed the Cohens were deeply implicated in the Rudolf Abel case. Photos of the Cohens and details of Morris Cohen's various aliases were sent to the Security Service. Bjoze, however, never visited Bermuda. In 1959 the Bureau asked MI5's help 'to trace the whereabouts of Cohen, who is involved in the Abel espionage case through an American rare book dealer, Jacobs,' who had communist sympathies. This resulted in what the Security Service described as a 'wild goose chase' and no further action was taken. Ironically, of course, unknown to both the FBI and MI5, Morris Cohen was at that very time already masquerading in London as an antiquarian bookseller.[34] However, 1959 did close with the FBI uncovering one more vital clue connecting the Cohens to Rudolf Abel. In November an old friend of Morris Cohen recalled having dinner at the Cohens' apartment in 1949 with 'Milt', a 'wealthy English businessman'. 'Milt' was clearly Abel.[35]

Despite the most intense efforts, at the start of January 1961 Alan Belmont and the FBI had no solid evidence of the movements and lives of Morris and Lona Cohen after they had fled New York in 1950. Then an urgent message arrived at the FBI's Washington headquarters from London's New Scotland Yard. The fingerprints of a couple called Peter and Helen Kroger, who had just been arrested for suspected Soviet espionage, matched those of Morris and Lona Cohen circulated almost three years before by the Bureau. Extra checks carried out immediately by Bureau fingerprint experts confirmed there was no mistake. Within days original reports from FBI files on the Cohens, with a request for the fullest possible briefings on the arrests and stunning discoveries at 45 Cranley Drive, were hand-delivered to MI5 headquarters in London.[36]

Part Three
Trial

6

'One of the most disgraceful cases to come before this court'

I

Breaking into Harry Houghton's cottage was a routine matter for Bill Colfer and the Dorset police. Colfer had waited all day on 7 January 1961 for the news of the first arrests to come through from London to police headquarters; around 5 p.m. he finally set off with seven officers through the chill winter gloom. A detective constable found an insecure window and, with the enviable agility of the young, soon clambered through and opened the front door.[1]

They spread out through the rooms and searched them methodically over the coming few days. The Admiralty material they found – such as three UDE test pamphlets about sonar equipment, stuffed with Houghton's customary lack of effort and imagination into the record compartment of his gramophone – was not particularly sensitive but should not have been in his cottage. Much more intriguing was other evidence of his espionage: a Swan Vesta matchbox with a primitive false bottom, containing a rectangle of paper with a sketch map of a road in west London's Notting Hill and markings for meetings with his controller; an Exacta 35mm camera and other camera equipment suitable for taking pictures of documents and developing them in the bathroom; and, on the pages of Houghton's desk diary, various markings relating to the dates of his meetings with Lonsdale. The garden shed held another poorly concealed secret: an empty tin of Snowcem paint containing a plastic bag, secured with two elastic bands, with £650 in £1 and ten-shilling notes.

Intent on securing important evidence, Bill Colfer and the police left Houghton's cottage and descended on Ethel Gee's address around 10.30 p.m. Understandably there was great alarm and distress when the local Dorset superintendent served the search warrant on

Gee's mother, aged eighty, and her uncle and bedridden aunt. The superintendent was forced to spend some time calming them before Colfer and two policemen could clatter upstairs to search Bunty Gee's bedroom. In one handbag they found in Gee's handwriting a list of subjects related to underwater detection equipment. UDE identified this as a 'comprehensive questionnaire which, if answered fully, would give the complete content of their work' on secret anti-submarine sonars. Details of secret research work at Portland were written on other pieces of paper discovered in Gee's handbags, and she was found to have considerable savings: £316 in cash and around £4,400 in bonds and shares.

Elwell liaised with the Admiralty to identify and assess all the material which Houghton and Gee had passed to Lonsdale on 7 January. The films were developed, prints made and taken down to Portland. Several hundred prints were from a secret Admiralty document called *Particulars of War Vessels* and from various Confidential Admiralty Fleet Orders. There were thirty-two photos of a technical drawing of the hull of Britain's first nuclear submarine, *Dreadnought*, and of various test pamphlets for sonars. The 35mm film found in Gee's shopping bag, when developed, proved to be completely blank, almost certainly because of Houghton's incompetence as a photographer. The four UDE documents physically transferred to Lonsdale by Houghton and Gee contained test results on aspects of different individual sonars. Only one had a security marking, the second lowest, of 'Confidential'. On 12 January the Admiralty vouchsafed to MI5 that it was 'preparing a considered . . . assessment of the damage which would have been caused' if the material given to Lonsdale had actually been transferred to 'an unfriendly foreign power'.

Meanwhile, MI5 were starting their own assessment, and seemed much more attuned than the Admiralty to the importance, for a hostile state, of acquiring intelligence about the revolutionary Type 2001 sonar being developed at Portland for the *Dreadnought*. According to one of the UDE technicians working on the project, within days of the arrests a team from MI5 were at the base quizzing them about how damaging the spying by Houghton and Gee might be for this sonar. Corralled into a local cinema for questioning, the project team were relatively sanguine, arguing that in many cases they were regularly changing the specifications and so any intelligence fed to the Russians by Houghton or Gee was outdated and so would have been of limited value. Later evidence suggests, however, that the technicians and scientists were over-optimistic.[2]

These questions were part of the review of protective security at UDE requested by the Admiralty immediately after the arrests. The Security Service sent down to Portland from Leconfield House a senior officer within a matter of days and his report was damning: 'Portland must be of great interest to a foreign power. In it is concentrated the Establishments concerned with underwater weapons and detection; . . . the "working up" base for all new construction vessels other than aircraft carriers [and] the centre for peacetime practice in the hunting and destroying of submarines.' The report continued, 'At a very rough estimate, there are about 100,000 Secret and Confidential documents in the Naval Base Establishment and this figure excludes papers . . . kept in a strongroom.' ('Confidential' were the least sensitive documents, followed by 'Secret' and 'Top Secret'.) The team working on the Type 2001 sonar at UDE was extensive: twenty-three scientists, twenty draughtsmen and twenty-four others. Yet, despite the signal importance for national security of the work carried out at Portland, only a total of fourteen employees had been positively vetted, nine of whom had access to atomic secrets. Over three pages the report made detailed recommendations for improvement: better security at entrances, better locks, and a rule that normal telephones should not be used for secret conversations.[3]

Lonsdale's tiny flat in London was quartered to find evidence in addition to the Ronson cigarette lighter and Chinese scroll. In the wardrobe was a deed box containing Lonsdale's false Canadian passport and birth certificate, a belt on a pair of trousers with two concealed zips holding 300 US dollars, and a tin of Yardley's talcum powder, which in two covert cavities hid a microdot reader and three miniature signal plans. Lonsdale's Bush radio was equipped with a lengthy aerial stretching all round the flat to ensure better reception for his messages from Moscow.[4]

The search of 45 Cranley Drive, where the Cohens had been masquerading as Peter and Helen Kroger with chutzpah for five years, had begun the instant after their arrest. Elwell, Winterborn and Peter Wright entered the house and straight away 'Winterborn got to work on some cigarette lighters' (as an MI5 memo phrased it) in the dining and living rooms. One of these had a wooden base and hiding place identical to the one owned by Lonsdale. Inside were OTPs and miniaturised radio transmission schedules. One of these was a duplicate of a schedule found in Lonsdale's lighter for the first and third Sundays of each month from April 1960 to January 1961. In the Krogers' bedroom they found Lonsdale's leather briefcase and on the floor by the door his microscope, suitable for viewing microdots. MI5 stayed in the bungalow until well after midnight. The Security Service laboratory at Saint Paul's opened early on Sunday morning to enlarge the microdots, and copy the coded signal typed on Lonsdale's typewriter and pages of manuscript in Russian in Lonsdale's handwriting. Winterborn and Wright passed all the next day scouring 45 Cranley Drive for a transmitter, but with no success.[5]

A search of the bungalow, initially by a team of six police officers followed by a more thorough one by MI5, continued for over a week: from the loft (where an aerial almost seventy-five feet long was strung across the rafters, four sheets of black-painted wood were fitted over the bathroom windows to turn it into a darkroom, and Exacta and Kodak Retinette cameras were found); to the bedrooms, where photographic developing equipment and a microdot reader – concealed in a carefully engineered hiding place in a tin of talcum powder – were discovered; to the sitting room where, hidden behind a bookcase, were two New Zealand passports in the names of Peter and Helen Kroger. The numerous books were examined page by page, and floorboards and insulating materials in the roof and upper floor were ripped out. One significant mystery remained: the jewel in the crown of the Cohens' espionage equipment, a transmitter, could not be found. Finally, in the kitchen the police pushed back the fridge and uncovered beneath the linoleum a trapdoor, which led down to a shallow cellar.

They searched it for two days using a length of electric flex almost fifty feet long with a light at the end (which had been discovered in the bedroom with no immediately obvious use), but uncovered nothing. Hugh Winterborn of MI5 and two colleagues therefore decided (in his words) on a 'further attack', armed with a mine detector. The detector proved 'valueless', so a physical search was started using an iron crowbar. It was hot and exhausting work exploring the cellar, square yard by square yard, crawling over and prodding the debris strewn everywhere by builders when they had partly de-molished party walls to install central heating several years before. Eventually, after two hours, about five feet from the trapdoor in the kitchen floor the crowbar struck something especially hard. They cleared away a pile of rubble. Beneath was a concrete slab. When levered out it proved to be four inches thick, and concealed beneath it was a specially made pit with smooth concrete sides about twenty inches by seven and eighteen inches deep. Inside were five packages wrapped in moisture-resistant polythene. The packages contained a transmitter with earphones and spare parts, a tape keying device to be connected to the wireless for 'squash' transmissions, microdot equipment, a miniature camera, and 6,000 US dollars in twenty-dollar bills.[6] Immediately a jubilant MI5 called in GCHQ to help assess the significance of the find. Cheltenham was unequivocal: the transmitter was to send 'flash' or 'burst' signals similar to ones they had seen previously from the same Russian 'stable' but of a newer and better design.

Once encoded by the keying equipment, it could send a message at a speed of around 270 words per minute in just a few seconds, making its interception almost impossible unless detection equip-ment was close by at the very moment of transmission. All the evidence, and especially the dates on the signal plans, suggested that the KGB had provided the Cohens with this novel transmitter as recently as April 1960. They would have used it only in special circumstances like emergencies, perhaps only on two or three occa-sions before their arrest, and not to send the product of espionage. It was no wonder that the one attempt by GCHQ in December 1960 to intercept radio transmissions from the bungalow ended in failure.[7] Meanwhile, the MI5 watchers who had been based at the OP in the Searches' house opposite for nine weeks needed to settle some debts. To thank their hosts for endless cups of tea and their hospitality, the two women agents gave the Searches a set of attractive fish knives. 'It was a matter of national security and we had to do our bit,' Bill Search declared later. 'We knew there was no reward or anything like that. We did it because it was our duty.'[8]

Immediately after the arrests MI5 had high hopes that Houghton and Gee would confess and reveal the full picture of what secrets they had passed to Lonsdale. The Security Service were to be dis-appointed. Superintendent Smith allowed Houghton and Gee a few minutes together early on the morning after the arrests. Shortly afterwards Houghton told Smith he wished to make a statement, saying: 'I am doing this to save Miss Gee because I have dragged her into this mess.' In the statement Houghton confirmed that he had passed documents to Lonsdale, but only because he believed him to be a Commander Alex Johnson of the US navy, who had asked him to supply him with information from Portland because 'the American authorities were anxious to know whether or not information supplied by the Americans [to the British] was being acted on'. Houghton said he 'used every power of persuasion' to compel Bunty Gee to photograph secret documents at Portland which he could transfer to Johnson 'to check on some of his bragging'. Gee, in other words, was 'an innocent pawn', as Houghton described her in a letter to relatives which he sent from Brixton prison and was copied to MI5.

Two days later Houghton, in Brixton prison, passed a message to Smith. He wanted a plea bargain. He told the superintendent he wanted to see photos of Russians and Poles attached to their UK embassies both past and present to identify those he had been in contact with and give evidence against the other prisoners, and in exchange not stand trial himself. Unsurprisingly, Smith refused, because the prosecution knew they already had powerful evidence against Lonsdale and the Krogers; they did not need Houghton's help. The same day, Gee, having been shown as part of her legal rights a copy of Houghton's statement, dictated one of her own corroborating Houghton's story about the mysterious US naval commander.[9]

At Leconfield House there was justified jubilation over the arrests. Roger Hollis instructed his subordinates to pass on his and the Home Secretary's congratulations to everyone involved. It was widely recognised that the new scientific techniques deployed in the case, like the use of 'Rafter' by GCHQ and the copying of the code pads, played a major role in the success.[10] One highlight for Hollis must have been his meeting with the Cabinet Secretary just over a week after the arrests, and Norman Brook's comment that 'he very much hoped we [MI5] would get some glory from this. He always felt we had been particularly hardly treated in the Burgess and Maclean case, where we should have been praised for a skilful detection but in fact only got criticised for allowing them to escape.'[11] Another was a personal letter of congratulation to MI5 from J. Edgar Hoover for 'discovering and exposing those who are dedicated to subverting our free world'.[12]

The decision to prosecute lay with the government's premier law officer, Sir Reginald Manningham-Buller (father of a future MI5 Director General, Eliza Manningham-Buller). Portly and bespectacled, the Attorney General was waspishly christened 'Bullying-Manner' by the journalist Bernard Levin, and according to some uncharitable contemporaries was not blessed with an outstanding legal brain. He did, however, attempt to reform legal education to allow interchange between barristers and solicitors, and was a sufficiently tough, capable and independent-minded lawyer to be appointed later lord chancellor in 1962 and a judge in England's highest court under Harold Wilson's Labour government in 1969. Devoted to his family, he was an intensely private man who asked for all his private papers to be burnt after his death to avoid not scandal but publicity. Educated at Eton about a decade after Harold Macmillan, he was a devoted follower of the Prime Minister and his 'one nation' brand of conservatism, and in turn was trusted by him.[13]

In private the Attorney General was genial and supportive, and impressed the Security Service. When he met MI5 for the first time to discuss the case on 12 January, its legal adviser described Manningham-Buller as 'most pleasant and helpful'. The Attorney General made clear that he wished to be actively involved with the case and would himself lead the prosecution at the Central Criminal Court.

With the overwhelming evidence of espionage collected in the aftermath of the arrests, there was no doubt that a criminal prosecution was appropriate and likely to be successful. None of the accused had confessed. This meant that a considerable amount of sensitive evidence would need to be called. The questions were how much of the case needed to be heard *in camera* and whether the government should fan the flames of publicity. In 1961 MI5 was a largely secret organisation and, remarkably, there was no reference to it at all in the first (1962) edition of Anthony Sampson's iconic and otherwise comprehensive *Anatomy of Britain* (the existence of the Security Service had only been publicly avowed in 1952). Hollis supported this hermetic blanket of secrecy, and was proud of the 'excellent arrangements with the press' which ensured that newspapers did not refer to the Security Service.[14] The role and evidence of the Security Service were, however, central to the success of the prosecution. This troublesome issue was to be resolved by calling the minimum number of MI5 officers to give evidence anonymously, and trusting the largely compliant media not to breach the mutually agreed screen of secrecy.

There was support from Macmillan, the Home Secretary, Rab Butler, and the Foreign Office for using the trial as a public relations coup to strike back at the Soviet Union in the Cold War. The KGB spy network uncovered in England could serve as an effective propaganda response to the shooting down of the American U-2 spy plane and arrest of its pilot, Gary Powers, on 1 May 1960.[15] In a show trial in Moscow, Powers had been convicted of espionage by the military division of the USSR's Supreme Court in August 1960 and sentenced to ten years' imprisonment. Powers languished in building number 2 of Vladimir Central Prison, about 150 miles east of Moscow, learning carpet-weaving from his cellmate, Zigurd Kruminsh, a Latvian political prisoner, to help pass the time.

In February 1961 his spy swap was in the distant future and Macmillan still harboured bitter thoughts about the U-2 incident. He had been preparing for two years for the major international summit in May 1960 in Paris with Khrushchev and Eisenhower. The shooting down of the U-2 that month had wrecked it. The usually languid Prime Minister had exploded in private, complaining of an 'undisciplined American general . . . blowing up the Summit Conference'. He took particular pride in the fact that the RAF's equivalent of the U-2 overflights, code-named 'Oldster', had all happened without incident.[16] In the months that followed the U-2 incident Macmillan must have resented the considerable amount of time he needed to devote to sorting out its consequences. These included new and secret rules to regulate US spying flights from the UK and related UK intelligence-gathering activities (such as flights along the borders of Warsaw Pact countries to gather electronic intelligence) to avoid the risk of a dangerous clash with the Soviet Union. The Prime Minister may also have been encouraged to embarrass Moscow by giving publicity to the Portland trial because of the recent and shocking discovery that the Russians had placed three microphones in some of the most sensitive locations in the British embassy in Moscow, the registry and cipher room. For twelve years until October 1959 they had been successfully intercepting information in, and to and from, the embassy up to and including 'Top Secret'. Manningham-Buller signed the official consent to prosecution nine days after the arrests.[17]

The Portland spy ring were arrested at a pivotal moment in the Cold War. High-definition photographs taken by the U-2 over Russia since its maiden flight in July 1956 had revealed to the Eisenhower administration that the USA possessed a gigantic superiority over the Soviets in terms of nuclear weapons, but because this information was so sensitive it could not be shared with the American people. During his 1960 presidential campaign, John Kennedy mercilessly exploited American paranoia about the alleged 'missile gap' with the USSR that Eisenhower had permitted to develop. It would clearly have been awkward too soon after becoming President in January 1961 for Kennedy to acknowledge no 'missile gap' existed. Instead he was forced to bide his time, taunted by a series of American setbacks and perceived Soviet successes: the debacle of the Bay of Pigs landing in Cuba in April; the USSR triumph the same month in placing the first man into orbit around Earth; a clumsily handled summit in June when Kennedy gave the impression to Kruschchev that America acquiesced to the permanent division of Berlin; culminating in the unopposed building of the Berlin Wall in August.[18] It was in this volatile international atmosphere that the arrest and trial of the Portland spy ring detonated.

One of the earliest and easiest – and, it proved, fortuitous – decisions the prosecution took was to spare Peggy Johnson in Malaya the agony of being a witness against her former husband, Harry Houghton. Mrs Johnson was already being harried with offers of money for interviews by journalists. The pressure was so intense that she and her husband were moved into the local RAF compound, and on the eve of the trial the Admiralty sent a personal message marked 'Exclusive emergency' to the commander-in-chief Far East asking him to contact his RAF colleague and make sure that Mrs Johnson did not reveal anything sensitive.[19]

More significantly, it was fortuitous because of a document which emerged from the personnel files at Portland for the first time on 17 January 1961, and was sent to the Security Service. It was a report by the probation officer there detailing how he had been approached for help by Mrs Houghton over her disintegrating marriage in July

1955. She told the probation officer that her husband was 'in touch with communist agents in London and is passing information on to them from the UDE files', and gave him a copy of a document which her husband had dropped in the house and was 'proof' of his subversive activities. The probation officer described Mrs Houghton as 'totally unbalanced' and her claims as 'really fantastic . . . either a case for the MI5 or [a mental] hospital', and he and his superior took no action.

The newly uncovered probation report was forwarded to MI5 by the Chief Constable of Dorset. He remarked acidly that if this information had been passed to the police back in 1955, there was little doubt in his mind that the case against Houghton and Gee 'could have been brought to a successful conclusion a long time ago'. It appeared that Mrs Houghton had repeated the allegations about Houghton's spying to several others at Portland – most remarkably perhaps to the local milkman, who reported to the Weymouth police as soon as the arrests were publicised. The only action the Security Service could take was immediately to forward the report to the Admiralty in London, aware of the embarrassment it would cause them if made public. The decision not to call the former Mrs Houghton as a witness in the forthcoming criminal trial would undoubtedly help spare the Admiralty's blushes.

The prosecution was planned on the basis that other possible evidence would also be omitted: evidence that would highlight the leading role of the Security Service in the case, cause embarrassment to the Admiralty, MI5, or the government, or would compromise secret intelligence. There would be no references to Mrs Houghton's statements nor to the probation officer's report, to the anti-Semitic letter which first piqued MI5's interest, nor in particular to MI5 breaking into Lonsdale's safety deposit box on 12 September, GCHQ's central role in decrypting Lonsdale's radio traffic, or Michał Goleniewski. These were not necessary for a successful prosecution. The court and so the public should be presented with a version of the case censored in the interests of national security.

The decision was taken that Lonsdale and the Krogers were to be charged in their false names. They had carried out their alleged espionage under this cover, and it was much simpler to prosecute them with these identities, and for the jury to decide whether or not they were guilty under these names. There was also a risk that the defence would argue that the jury would be prejudiced when hearing the evidence against the Krogers if they knew about their flight from the USA in 1950 and allegations that they had spied for the USSR during and after the war. Although MI5 knew the real identity of the Krogers many weeks before the committal hearing, it was never made public at that time. The Security Service were apprehensive about the forthcoming criminal trial being compromised when photographs of the Krogers were inevitably to be published in the USA and they were identified as the Cohens. In Washington the FBI agreed that if this happened and they were approached, the Bureau would state that they had circulated fingerprints of the Cohens to New Scotland Yard three years before and they were identical with the Krogers' – but strictly nothing more. They would certainly not offer any help to journalists who visited the USA on a 'fishing expedition'.[20]

Meanwhile, Elwell and Colfer pressed ahead with essential follow-up enquiries. The names and addresses of all contacts found in the papers of Lonsdale and the Krogers were assiduously sifted, with MI6, the RCMP, the Americans and even the Australian intelligence service asked to help. Photos of the three illegals were shown to Russian defectors for possible recognition. The celebrated 1946 defector, Igor Gouzenko, had no knowledge of them; nor did the Petrovs, whose defection in Australia in 1954 had created headlines around the world. Their comments confirmed MI5's assessment: 'The Petrovs believe that these persons were members of an Illegal Residency in the UK and that their infiltration and establishment in the UK would have organised by the Illegals Directorate in Moscow. This directorate is segregated from the State Security Service and the Petrovs would therefore not have been likely to see them in the precincts of the KGB building.'[21] Interviews with, and statements from, the motley assortment of girlfriends of Houghton who figured in the investigation confirmed that none was involved with his espionage. A fellow student of Lonsdale at SOAS verified that the illegal visited Ruislip from as early as 1955.

Elwell examined the passports of Lonsdale and the Krogers minutely to draw up a schedule of their arrival in Britain and their international movements afterwards, and asked MI6 to confirm or supplement it from their contacts abroad (the Danish intelligence service, for example, confirmed that when Lonsdale disappeared from London on 27 August 1960 he travelled to Denmark before moving on to Moscow). Other allied security services around the world were enlisted to make enquiries about the Krogers and Lonsdale, in Austria, Belgium, France, Holland, Italy, Japan, Portugal, Spain, Switzerland and West Germany.[22]

Liaison with the RCMP confirmed that Lonsdale's Canadian passport was indeed false, and contacts with sister intelligence agencies in Western Europe through MI6 showed how Lonsdale, after his arrival in the UK in March 1955, had made long trips abroad to numerous European countries during the summer months of the following years. Such evidence helped confirm that Lonsdale was a Russian illegal, but Elwell knew that it would be an immense challenge to establish his real identity. A helpful start was irrefutable evidence that the man held in Brixton prison not only held a fake Canadian passport but that he could not possibly be the real Gordon Lonsdale.

The RCMP had continued to investigate the background of the real Gordon Lonsdale. On 16 January they reported to MI5 that they had finally tracked down and interviewed his father, Emmanuel Lonsdale, a man with Native American ancestors, who confirmed that his baby son had been circumcised. Elwell passed this information to Superintendent Smith, who reported: 'I saw the doctor at HM Prison, Brixton and at my request Lonsdale was medically examined . . . on Monday, 30.1.1961, and it was found that he *has not* been circumcised. It is abundantly clear in view of the foregoing that the accused Lonsdale is not identical with the son of Emmanuel Lonsdale.'

Around the same time, FBI agents visited a small rural community just south of Seattle in Washington State. There they interviewed the sister of Lonsdale's mother. She confirmed that after his mother, Olga, separated from her husband in the early 1930s, she emigrated to the USSR with her new Finnish partner and young son. In the weeks that followed, the Bureau and the RCMP located other relatives and friends of the Lonsdale family who told the same story.[23]

Since they were clearly false, Elwell began an investigation into the Cohens' New Zealand passports issued in the names of Peter and Helen Kroger and how they had been obtained illegally. The files were swiftly retrieved from the archives. They showed that, in the name of Peter Kroger, Morris Cohen had sent the request for the passports from a small hotel in Austria to the New Zealand consulate in Paris in April 1954, on the basis of an existing British 'family' passport in the name of Kroger, his New Zealand and his wife's Canadian birth certificates, and a marriage certificate. Cohen asked for the return of this British passport when his application was complete. Elwell's enquiries disclosed why. The British passport was one of a batch of unused and blank passports sent in 1947 to the British consul in Shanghai, who was ordered to destroy them all in 1951. This unused passport had clearly escaped destruction and fallen into the hands of the KGB, to be doctored and reissued as a forgery to conceal the Cohens' true identity and nationality. The passport was filled in with a number which, when investigated by the Passport Office and MI5, proved also to be that of a genuine British passport issued to 'a very respectable lady' who had been in the UK throughout this period. False too were the Krogers' birth and marriage certificates. Research in New Zealand discovered that an unknown person had obtained a birth certificate with the same number as Peter Kroger's two months before Kroger first wrote to the New Zealand consul in Paris. The KGB forger had, however, made a tiny but significant slip: he had added a letter in front of the number, which was not New Zealand practice. There was another slightly puzzling anomaly on the file: the first secretary in the New Zealand consulate who dealt with the application did not sign any of the letters. Elwell told the New Zealand High Commission in London that he thought their 'consul [in Paris] could scarcely be blamed for not having smelt a rat when confronted by the documents produced by the Krogers'. Always diligent in following up leads, Elwell was to dig further into the mystery of the New Zealand passports after the criminal trial, with unexpected results.[24]

The first stage in the prosecution was the committal proceedings, when the prosecution was required to set out before a court of magistrates that there was a case to answer. The defence would only be presented later at the trial of the accused at the Central Criminal Court, known as the Old Bailey. The committal proceedings opened in the cramped courtroom number 1 at Bow Street Magistrates' Court on 7 February 1961 – normally the setting only for summary trials of petty thieves, drunks and prostitutes. Every night for the preceding week carpenters worked into the small hours erecting a new floor in the public gallery with raised benches so that journalists could see what was happening in the well of the court. Only one important country outside the UK was not represented: Russia. TASS, the official Russian news agency, did not send a single representative to cover the proceedings, despite its sizeable staff in London.

When the hearing opened, in a demonstration of the significance of the case, the Attorney General himself presented the prosecution case. He underlined that the prosecution was sure that Lonsdale was a Russian national, and that the spy ring was stealing secrets for Russia. Lonsdale's face was mostly expressionless, only displaying some interest when the A4 watchers gave evidence of their surveillance; Peter and Helen Kroger were animated, waving to friends in the court, and followed proceedings closely; Houghton occasionally muttered comments and appeared to enjoy being the centre of attention; Gee was pensive and monosyllabic. The defendants pleaded not guilty and reserved their defence. On 10 February all five were committed for trial.[25]

After the searches and committal hearing were complete, 45 Cranley Drive was released to the Krogers' lawyers to make arrangements to sell the bungalow and dispose of its contents. It was then that another extraordinary discovery was made. Michael Bowers was the business associate of Lonsdale who was employed to make an inventory of the Krogers' belongings. He had been searching for several hours when he found a brown leather writing case in a drawer of the dressing table of the Krogers' bedroom. There was a shield embossed on the front including the word 'Bruxelles'. Bowers said later that because he 'regarded everything at 45 Cranley Drive with suspicion I decided to examine the case more carefully'. He therefore took it home, but forgot about it for a day or two because of pressure of work:

> When I did finally examine it, the fact that a distinct crinkling sound was produced when I rubbed my thumb along the outer surface made me investigate further. I imagined it contained bank notes. As the writing case appeared to be new, and not wishing to do any unnecessary damage, I made a small incision under the writing pad. The razor blade which I was using snagged what appeared to be a white envelope, and through the opening I could see part of a coat of arms. As my curiosity was

now thoroughly aroused I enlarged the cut and found a false compartment containing a white envelope made of paper which crinkled. Inside I found a Canadian passport in favour of a Mr Thomas James Wilson, but with a photograph . . . of Mr P.J. Kroger . . . I found a similar false compartment in the other side of the writing case and in it another Canadian passport. This one was in favour of a Miss Mary Jane Smith, but with a photograph of Mrs P.J. Kroger.

Inserted into the false Wilson identity document was a piece of paper setting out details of the passport, and, under the heading 'Meaning of stamps', giving an explanation of how Wilson came to Montreal in Canada by plane on 23 July 1956 via Holland and Belgium, before saying: 'Then left Canada for England any time. Note that the immigrant authorities in Canada do not stamp Canadian passports when leaving the country. In England they stamp Canadian passports neither on arrival nor when leaving the country.' A similar explanatory note was attached to the Smith passport. Bowers also discovered 4,000 US dollars in a hollow book-end, and in another hiding place a blank British passport, waiting to be completed with the details of whoever planned to use it. The estate agent took the two obviously false 'escape' passports and other discoveries to the police on 15 February. He at first refused to make a statement, but was finally persuaded to do so by Superintendent Smith just three days before the criminal trial opened.[26]

Another intriguing item emerged which had escaped the police and Security Service searches after the arrests. As the trial approached, Houghton was anxious that his solicitor should search his cottage for a half-page cut out from a pocketbook in connection with a statement he was thinking of making at the trial. The lawyer finally found the missing paper inside the cover of a vinyl record. On it was written the address of a flat in Lancaster Road, Notting Hill, a short distance from the part of Ladbroke Grove that featured on a sketch map found earlier in the matchbox in Houghton's cottage. What was astonishing was who lived there: a junior Russian diplomat named Vasili Dozhdalev. When Special Branch informed Elwell of the discovery on 3 March, he drew the obvious conclusion: 'the occupant of the flat there was Houghton's controller before he was taken over by Lonsdale'. Dozhdalev, it turned out, had suddenly returned to Moscow only a few days before, on 22 February. Before the discovery of that scrap of paper the Security Service did not know that Dozhdalev was a KGB officer and that he was linked to Houghton. Houghton had broken all the established rules of trade-craft by keeping a copy of the address of the KGB man without any attempt at encoding.[27]

The prosecution understood that if the jury found the defendants guilty, there would need to be openness about what MI5 knew about the Krogers and Lonsdale before the judge sentenced them. In the weeks before the trial, Elwell worked with Superintendent Smith to prepare a lengthy statement summarising information accumu-lated about the Krogers and Lonsdale by the Security Service, in a form which MI5 was content to be made public.[28] This document underlined how extensive liaison was with both the FBI and the RCMP about Lonsdale – with the FBI, for example, checking out the schools in California which Lonsdale stated he had attended on his SOAS application form – and about the Krogers, confirming details of the Canadian 'escape passports' found at their address and how the RCMP laboratory analysed them and said they were 'complete forgeries'. The FBI began an extensive trawl through Peter Kroger's bibliophile contacts in America to sieve out any who might have communist connections.

Before the trial, MI5 and the police pieced together further details of the lives of the Krogers. They learnt how Peter Kroger developed his 'legend' as a bookdealer from Canada who had moved to Britain to do business because of his health (he said he suffered from boils on his backside); slowly built up his purchases, partly by outbidding others at auction; was popular in the book trade, entertaining contacts generously at Cranley Drive; and even played in the Bibliomites' annual cricket match, wielding his willow like a baseball bat and trying to hit home runs. Kroger made no attempt to hide his North American origins, dressing in a pale-blue suit with wide lapels, a white nylon shirt and basket-weave brown shoes.[29]

Whispers were starting to circulate around the secret corridors of Leconfield House and Whitehall about how much the Security Service knew about Houghton before their investigation began at the end of April 1960. To sift rumour from fact, the head of counter-espionage, Martin Furnival Jones, asked Arthur Martin to produce a report tracing the information about Houghton provided to the

Admiralty in London, UDE at Portland and finally to MI5. When laid out in chronological order it made uncomfortable reading. 'FJ' commented that '[looking]at it dispassionately now I think it is clear that we ought to have carried out some investigation in 1956. If we had done so there is a fair chance that we would have unearthed Houghton's espionage . . . some four year earlier. The consolation is that in the event we have hurt the RIS more' by capturing Lonsdale and the Krogers. Roger Hollis read these notes three days before the trial started, no doubt already thinking ahead about how best to counter any future criticism of the Security Service.[30]

Also hidden from the public gaze, there was burgeoning disquiet in the Admiralty in particular and in Macmillan's government in general about how the case would be reported. The Admiralty put repeated pressure on Peggy Johnson in Malaya not to reveal to journalists her suspicions about Houghton and that she had reported them to the authorities.[31] There were justifiable fears that the press would lambast the security arrangements at naval and other military sites, and these articles would provoke an adverse reaction – especially in America. Lord Carrington, the First Lord of the Admiralty, sent a warning note to Macmillan on 9 March, a few days before the start of the trial, copied to the Foreign Secretary, Alec Douglas-Home, and the minister of defence, Harold Watkinson.

In fact, the Prime Minister was already well aware of developments in the USA. One of the most important priorities for him at that time was to forge a bond with the glamorous new president, John F. Kennedy, and seek his support over Britain's first application to join the European Economic Community and various nuclear weapons issues. Macmillan had been invited to America for significant talks with Kennedy in April, and at the start of March he had received a personal note from the minister of state for foreign affairs, David Ormsby-Gore, who was a friend of the president and had private discussions with him in Washington. Ormsby-Gore said that Kennedy was looking forward to his talks with Macmillan, 'but there are some hazards along the otherwise broad and inviting fairway'. One bunker was that the approaching Portland spy trial had aroused 'Congressional suspicions about our security'. Ormsby-Gore gave the president an assurance that 'we know of no evidence that any classified information regarding the nuclear propulsion unit of the British [nuclear] submarine has been compromised'. But, the minister continued, it was clear that 'some damage has been done', and Kennedy suggested that when he met Macmillan they should discuss 'existing security screening procedures' so he could reassure Congress and others that American secrets were safe when passed to the British.[32]

In the weeks that followed, concerns about the reaction to the trial by Kennedy and in America in general – perhaps restricting future defence collaboration – had wide repercussions. One of the most significant was a desire to manage much more aggressively the publicity generated by the criminal trial. Harold Watkinson wrote to the Prime Minister, warning that 'the exceedingly unfortunate disclosures' to be made during the spy trial 'would not be a very good augury' for Macmillan's USA visit. As a result, Carrington spoke to the Attorney General about minimising the 'damaging information' to be used in the prosecution. Manningham-Buller confirmed that he already had 'this very much in mind'.[33]

II

The scene at 10.30 a.m. on Monday, 13 March 1961 in courtroom number one of the Old Bailey was closer to the Victorian era than twentieth-century Britain. Three staccato raps were heard from behind the oak door on the dais. A costumed usher jumped to his feet and declaimed, 'Silence!' The hubbub in the room, panelled in light oak and crammed with newspaper and media correspondents and thirty curious members of the public, faded to a hush as the Lord Chief Justice of England, Lord Hubert Parker, entered. He was accompanied by the aldermen and sheriffs of the City of London in their full robes and regalia of office. Lord Parker processed to his tall red-leather chair, where he paused for a moment to glance around the courtroom before taking his seat. His presence presiding over the trial was an indication of its importance. The five defendants were led up from the basement to the glass-walled dock, the Krogers and Lonsdale stood together clutching newspapers and writing materials, and Houghton next to Gee. Harry Frederick Houghton, Ethel Elizabeth Gee, Gordon Arnold Lonsdale and Peter John and Helen Joyce Kroger were charged with 'conspiracy to communicate information in contravention of Section 1 of the Official Secrets Act, 1911'.

The particulars of the offence were that the five defendants 'between the 14th of April 1960 and the 7th of January 1961 conspired together and with other persons unknown for a purpose prejudicial to the safety or interests of the State to communicate to other persons information which might be directly or indirectly useful to an enemy'; 14 April 1960 was chosen because it was the starting date of the radio signal plan discovered in the Krogers' bungalow, and would minimise the need for evidence of earlier events which would be inconvenient and irksome to the Admiralty and others. Altogether there were to be thirty-eight witnesses for the prosecution, nineteen of whom were anonymous (described only as Witness 'A' or 'B', for example) because they worked either for the Security Service or GCHQ. Manningham-Buller – imposing and florid-faced – led the prosecution, with two barristers, one of whom was Mervyn Griffith-Jones, who had a reputation as a conservative reactionary and had recently appeared at the Old Bailey leading the prosecution in the celebrated *Lady Chatterley's Lover* case.[34] Fronting the defence of the Krogers was Victor Durand. He enjoyed a powerful court presence and understood well how a defendant's appearance influenced a jury (once asking a gangland client from London's East End charged with theft-related offences to remove his glittering gold rings before the trial). Lonsdale's principal barrister was William 'Barry' Hudson, later to become a distinguished defence advocate, and who was to play an unexpected but significant role in his client's life after the trial. Houghton and Gee also paid for their own defence (in their case through selling their stories to newspapers) but were not represented by lawyers of any distinction. Watching proceedings were a clutch of anonymous representatives from Britain's intelligence agencies involved with the case, including of course Charles Elwell, but also his Security Service bosses Roger Hollis, Malcolm Cumming and Arthur Martin, and Arthur Bonsall from GCHQ.[35] Representatives of the FBI and the RCMP in London also attended.

Before later reforms, it was possible in the England of 1960 for potential jurors to be challenged by barristers without reason, but unlike in the USA they could not be cross-examined for suitability. A number of women, and various candidates who appeared to be from the higher social strata, were excluded as a result. After the inevitable challenges, the jury sworn in for the Portland trial consisted only of twelve men of middle age and respectful demeanour.

The Attorney General began by clarifying that the defendants were not charged with the act of espionage itself but with a conspiracy, an agreement, to carry out the criminal act of espionage. And because people 'who engage in a conspiracy do not do it in public', the 'proof of a conspiracy is generally a matter of inference drawn from the acts of the parties accused'. He underlined that for the defendants to be found guilty it was not necessary for each of them to know each other (the prosecution was well aware that they had no evidence that Houghton and Gee ever met, or even knew of the existence of, the Krogers). More observant journalists noted that Manningham-Buller was less expansive and cautious in his opening remarks than during the committal proceedings – when, for example, he alleged without hesitation that Lonsdale was Russian. At the trial he was more circumspect, not wishing to draw attention to the detail of the Security Service's intelligence on the spy ring. One of the few moments when Lonsdale showed emotion during the trial was when the Attorney General read out extracts from letters to and from his family, presumed to be in Russia, full of gossip about his two children. The manuscript letter in Russian from Lonsdale found in Helen Kroger's handbag was in response to one contained in a microdot. MI5 had the correspondence translated into English and Elwell had already been poring over it to extract clues which might reveal Lonsdale's true identity.

To protect the anonymity of all trial witnesses from the intelligence services, they were allotted a special room in the Judges' Corridor. They entered the courtroom by way of the bench where Lord Parker sat, gave their testimony, and discreetly vanished. The first two witnesses were MI5 watchers who described their surveillance of Houghton and Gee when they met Lonsdale near the Old Vic on 9 July 1960. On the bench in the little park opposite the famous theatre, for about ten minutes Lonsdale was left alone chatting to Gee, who 'appeared to be talking twenty to the dozen'. On the second day the defence tried to undermine the credibility of the two watchers who overheard the conversation between Houghton and Lonsdale in Steve's Café by questioning how far they really could hear fragments of their discussions against the background noise. They had little apparent success. Witnesses gave evidence about Lonsdale depositing various items in the bank in Great Portland Street at the end of

August, and how Special Branch had accessed the deposit box with an official search warrant in September. There was no reference, of course, to MI5 secretly examining the contents earlier that month. Further watchers in the witness box, including a 'Miss K', according to one observer 'looking like a secretary in a smart blue coat', told the jury the story of the later surveillance operations which led MI5 to the Krogers' bungalow in Ruislip and how they kept watch from a neighbour's house.

Superintendent George Smith gave evidence about the arrests at the Old Vic and Ruislip. Ever the showman, he demonstrated to the jury how to open the secret cavities in the Chinese scrolls discovered in Lonsdale's flat. Smith presented other ingenious items associated with Lonsdale's espionage, showing the jury the concealed zips on Lonsdale's money belt and shaking some talc from the fake Yardley's powder tin. The superintendent continued with details of how he took down Houghton's statement, including the story that he was approached first by Lonsdale pretending to be an assistant American naval attaché. Ethel Gee's barrister tried rather desperately to develop the idea that it was reasonable for his client to assume Lonsdale was American. He asked Smith:

'And he [Lonsdale] did, indeed, have certain transatlantic characteristics, such as chewing gum. Have you noticed that?'
A: 'I found some chewing gum in his pocket, yes . . .'
Q: 'Would you say even by the cut of his suit he would appear to have a sort of transatlantic manner about him?'
A: 'No, I couldn't agree on that.'

One amusing moment occurred when Gee's counsel asked Smith: 'She seems to have led a rather dull sort of life?' The superintendent smiled, but before he could reply Lord Parker intervened, 'How can the officer say that?', to suppressed titters around the courtroom.

Day three continued with evidence from a navy expert about the radio equipment found in the Krogers' cellar: a veil was cast over GCHQ's crucial role throughout the trial. He confirmed Cheltenham's conclusions reached weeks before. The transmitter was not made in Britain and its 'flash' transmissions could easily reach Moscow. Two anonymous MI5 scientists gave evidence about, for example, how the typeface of the Royal typewriter in Lonsdale's flat matched the series of five-digit numbers typed on the piece of paper discovered in Helen Kroger's handbag and another found in Peter Kroger's study in a book auction catalogue, and about the equipment used to read and create microdots. They confirmed how bungling Houghton was when taking photographs with his miniature camera. When cross-examined about the poor quality of the contact prints developed by MI5 from the film found in Gee's shopping bag, one of the witnesses agreed that good microdots could not be made from these unclear negatives: 'I take your point; yes, I don't think one would be very successful with these.'

Day three ended with evidence about Houghton and Gee's finances. Neither had a satisfactory explanation of the wealth they had accumulated, which jarred with their paltry annual salaries. Gee, for example, earned about £590 a year but made cash payments into her bank account or bought bonds worth £1,322 from May 1957 onwards. A language expert explained that the Russian words on the signal pads found in Lonsdale's Yardley's talc tin were radio call-signs, such as the name of a Russian river (the Lena) or town (Azoff); and one heading meant 'Transmissions on order of [Moscow] Centre'.

The prosecution case concluded on the fourth day.[36] It began with a self-confident naval interpreter ('I know sixteen languages and I am acquainted with naval terminology in twelve') giving his assessment of the questionnaire found in Gee's bedroom. His view was that it was not written by someone with knowledge of British naval language and terminology and whose native language was English. For example, the word 'transducers' had been incorrectly translated as 'radiators' (Russian being the only language where the literal translation of transducer is radiator), and he said that the phrase 'in conditions of high speed cruising' was not 'an expression which we in the British navy would use . . . We should say, "when proceeding at high speed" or "at high speeds".' The hapless security officer at Portland confirmed how lax security was in the drawing office where Gee worked, and how easily she would have had access to drawings of HMS *Dreadnought* and the test pamphlets.

Seven anonymous GCHQ wireless operators gave details of an operation to establish the origin of messages in Morse transmitted from abroad on 9 and 18 January 1961 at frequencies and times set out in the signal plans found in the homes of Lonsdale and the

Krogers. On each occasion bearings were taken on the messages from three secret listening stations in England and then triangulated to the location from which the messages were transmitted. The conclusion, reported the 'communications specialist in headquarters directing the stations', was that, although there was a range of error of around 650 miles, beyond doubt the signals' source was 'within the territory of the USSR'.

The witness put forward by the Admiralty to convince the jury and the media of the damage wreaked by the spy ring was Captain George Symonds, Director of the navy's Undersurface Warfare Division. He stressed the 'marked value' which the test pamphlets discovered in Gee's handbag and Houghton's house would have for a potential enemy, who 'would be able to make use of them in a way which would be highly damaging to our forces in war'. As for the questionnaire found at Gee's house:

> 'The answers to those questions . . . would be of extreme value to a potential enemy in that [they] would paint a picture of all our current anti-submarine equipment and our future anti-submarine equipment . . .'
>
> Q: 'Would you put that as a matter of the highest value?'
> A: 'I would, Sir.'

Symonds gave evidence about the thirty-one photos of the drawing of HMS *Dreadnought*. This drawing, the captain stated, contained information about 'the British designed anti-submarine [sonar] set and would be of undoubted value to an enemy'. In an amusing interlude, Houghton's barrister succeeded, however, in showing that some of the Admiralty papers found at Houghton's house would be of little value to the Soviets:

> Q: 'And how many of those eight [papers] . . . are of value to a potential enemy?'
> A: 'All of them, in varying degrees.'
> Q: 'One of them is concerned with medical treatment, is it not? May I mention what that treatment is in open court?'
> A: 'Certainly.'
> Q: 'It is concerned, isn't it, with the treatment of venereal disease?'
> A: 'It is.'

Q: 'Do you regard that as being of value to a potential enemy?'
A: 'It shows, to a certain extent what is going on in the navy.'

Ethel Gee was escorted to the witness box by a prison matron. She gave evidence for five hours over two days. Gee had a rather high-pitched voice and spoke softly with a Dorset burr, so several times the judge needed to ask the shorthand writer to read back her replies. She gripped one of the wooden posts supporting the canopy over the witness box or rested against the ledge at the front as she answered counsels' questions. Her evidence suggested someone who was not gushing or sentimental, with excellent control of her emotions – but also endured loneliness, sharing a tiny house with her elderly mother and uncle and bedridden aunt:

Q: 'Would it be right to say that you are particularly fond of your mother, of course, and of your uncle?'
A: 'Well, I have no one else.'
Q: 'Pardon?'
A: 'I have no one else.'

Gee had no hobbies or membership of clubs or societies because she had no time: there were always chores to be done at home after work. If she was going to vote, she would vote Liberal. She had no active political interests. She had first known Houghton in around 1953 when he started work at UDE in an office close by, but only began to see him separately early in 1958.

The picture which Gee's barrister painted was of a woman devoted to her elderly relatives, and of simple tastes. She said she was fond of Houghton, spent time with him, her feelings were reciprocated and they would have married before 1961 if her domestic circumstances had been different. Elwell and his MI5 colleagues in the courtroom must have smiled inwardly at this point, knowing of Houghton's bevy of girlfriends and his disparaging comments about Gee made in intercepted phone calls only shortly before the arrests. From 1958 she worked near the Portland strongroom, where sensitive drawings were kept, but Gee was keen to demote its significance, describing it as 'largely a dumping ground for old drawings . . . [and] . . . spare copies of the test pamphlets'. People were going in and out all the time. 'You never knew who you were going to find in there.' At which point a wag at the back of the court-room murmured, 'Khrushchev', which caused a ripple of suppressed guffaws.

Gee denied she gained any technical knowledge about the test pamphlets or drawings she dealt with, and sought to give the impression that she was not interested in her work and not particularly intelligent. She echoed Houghton's story that before the arrests she had always understood Gordon Lonsdale to be Alex Johnson, an American assistant naval attaché, to whom she was introduced for the first time outside the Old Vic on 9 July 1960. Her account was stilted and unconvincing, claiming that there was no purpose for that meeting other than for her to be introduced to Johnson. As for the 10 December meeting in London, Gee said they were going to meet Johnson 'to pick up a camera'. She saw nothing sinister in this: 'I know Americans are very fond of cameras. It seemed a likely thing.' Inconsistent evidence from Gee followed, culminating in her being persuaded by Johnson and Houghton to obtain some copies of test pamphlets to show the American. 'At the time I didn't feel I was doing wrong. I thought it was foolish. I see now that it was very wrong.'

More unconvincing testimony emerged when Gee described events on 7 January 1961. She said she had taken the seven test pamphlets on the spur of the moment just before leaving work the evening before, because 'perhaps they would be of some use' to Houghton and Johnson. She denied she ever knew the contents of her shopping bag on the way up to London because Houghton had borrowed it to carry a comic annual and a tin of tongue. Gee confirmed her shock when arrested, commenting that at first she thought Superintendent Smith 'appeared a gentleman . . . mixed up with what appeared to be a lot of Teddy Boys'.

Elwell and his fellow Security Service officers watched inscrutably as she gave her testimony, aware of evidence (and particularly the highly sensitive intercepts) that could prove that her unconvincing narrative was riddled with lies but reconciled to the fact it must remain secret. Although confident the jury would not believe her, they were relying on the Attorney General's cross-examination to demolish any remaining credibility Gee's testimony possessed. They were not disappointed. Manningham-Buller rose to his feet, adjusted his robes and quickly exposed some key inconsistencies and untruths, in particular about her awareness of the results of her actions:

> Q [Attorney General]: '. . . the arrangement . . . being [after the meeting with Lonsdale in London on 10 December 1960] . . . that you were to produce test pamphlets, Houghton was to photograph them and Houghton was to give the photographs to Lonsdale?'
> A: 'That is what it was . . .'
> Q: 'You meant Lonsdale to see the contents of those test pamphlets?'
> A: 'Well, either to see them or learn what they contained.'
> Q: 'You meant Lonsdale to learn what they contained?'
> A: 'Yes.'

When cautioned, Gee had said, 'I've done nothing wrong.' At first in the witness box she commented that she did not feel she had 'done anything criminal'. Manningham-Buller took her through her signed declaration of the Official Secrets Act dated 30 October 1950; Gee confirmed that she had no authority to divulge the test pamphlets to Houghton, and that it was a breach of her duty to take the pamphlets out at weekends. The judge intervened, clearly not convinced by Gee's total lack of interest in the seven test pamphlets she had taken from UDE on the Friday night before the arrests and passed to Houghton:

> Q [Lord Chief Justice]: 'Well, I do not understand. You had done, according to your own story, something wrong, something which you say you have never done before, in taking seven test pamphlets and giving them to Houghton the night before?'
> A: 'Yes.'
> Q: 'On the Saturday when you met him, did you not ask him: "Well, were they of any use, those test pamphlets?"'
> A: 'No.'
> Q: '"What have you done with them?"'
> A: 'No.'
> Q: '"Are they safe?"'
> A: 'No.'
> Q: '"Am I going to get them back?"'

A: 'No. I just trusted him . . . I could never believe that they would
be of use to him.'
Q: 'But you gave them to him?'
A: 'Yes, and I thought it was really foolish.'
LCJ: 'Very well. That is your answer.'

Gee maintained her unruffled self-control throughout, and at the
end of her testimony returned calmly to the dock to rejoin Houghton, who would give evidence next.

Behind the scenes during the first week of the trial there was a
flurry of activity to manage its public relations fallout. A meeting
was hastily convened on day three of the trial with the minister
of defence, Harold Watkinson, Lord Carrington and Hollis. They
agreed it was impossible to influence the press not to print hostile
articles, and nor should false stories be fed to the press (for example,
that Houghton had been under surveillance since 1956). Instead, two
versions of a statement would be prepared: one 'a true statement of
the case as it could be best presented'; and a 'rather freer version . . .
which would present the government in a favourable light'. At a
reconvened meeting the following day the minister of defence was
clear: 'it was better to stick to the truth'. It was agreed that after the
trial the press should have 'maximum facilities' to take photos of
the spy equipment, but details of any action to be taken as a result
should be delayed until after the government made a statement in
Parliament. Meanwhile the Admiralty had an 'off-the-record' discussion with the head crime reporter at the *Daily Mail*, spilling 'a
bit of information to find out what he knows, which is everything:
the whole history of the Krogers etc. . . . The attack will spread over
everywhere, FBI, MI5, RN, RAF and Army . . . His strong advice
was that the Admiralty should keep as silent as possible.'[37]

III

Back at the Old Bailey, there was hushed expectation as Houghton
bustled into the witness box. Dressed in a tweed suit, he was a short
man who – in sharp contrast to Gee – spoke up confidently and
loudly, and at first he appeared at ease, almost lounging, with one
hand in his pocket. Although he despised Houghton as a tawdry

traitor motivated by money, Elwell straightened himself on his green leather chair to be alert to analyse Houghton's testimony. There was clear evidence from the Polish defector Goleniewski that Houghton had been spying for Russia for a decade from the early 1950s, but this was much too sensitive to be revealed in court. Instead of confessing as Elwell hoped, Houghton had invented the fanciful story about Gordon Lonsdale being an American naval attaché and Elwell expected Houghton to embellish his tale from the witness box. Houghton did not fall short. As one crime correspondent wrote: 'Here was the spy story de-luxe with all the trimmings – the threats, the codes, foreign names, the mystery woman, secret meeting places, and the passing of the vital information.'[38]

Houghton began by rattling through his career in the navy, which included periods serving during the war on the perilous Malta and Murmansk convoys, and ended in him being promoted to a master-at-arms, the highest non-officer rank in the Royal Navy. According to Houghton his troubles began during his posting to Warsaw in 1951. It was there that he said he started an affair with a Polish girl called Christina, who visited him 'frequently' at his flat when his wife was back in England, despite the risk of her being spotted by the Polish Security Service. While in Poland he engaged in black-market activity, buying items in England and selling them at extortionate profit in Poland. After his return to England, Houghton claimed he received a mysterious phone call at work in early 1957 from a man who said he was a friend of Christina. They agreed to meet outside the Dulwich Art Gallery in south-east London one Sunday lunchtime. Houghton said he thought the man was a Pole, who immediately asked Houghton for information about naval matters and made a 'serious threat which could be carried out in this country' if Houghton reported the approach to the police. The Admiralty clerk declared he had 'no alternative' but to agree to meet the man again when he sent him the signal, a Hoover vacuum cleaner brochure. He was then to come up to London. At the end of 1957 he received the brochure and had a rendezvous with the man at the Toby Jug pub on the A3 Kingston bypass, but Houghton claimed he only passed him some old copies of a local newspaper containing 'a terrific amount of unclassified information about the Fleet'. By now warming to his theme, Houghton became almost garrulous, as though ensconced on a bar stool in his local pub. His genuine voice emerged, the ex-navy man, the teller of tall tales. On several occasions his own barrister was forced to intervene and cut him short.

Understandably, the mysterious man was not satisfied with the harmless titbits of information which Houghton was giving him and threatened to beat him up.

> H: 'I told him that such things can't happen in this country, I could get police protection. He said, "Do you remember a man by the name of Petrov [an important KGB defector in Australia in 1954]?" I said, "It strikes a chord." He said, "He was in the Soviet embassy in Australia. He went over to the West", he said, "but the Australians can't protect him . . ." He said, "Well, the Australian government are giving him £2,000 a year and protecting him, but yet he is in fear of his life." He said, "Do you know that that man now is a drooling alcoholic? And as soon as he goes the Australian government will be pleased about it . . ."'
> Q: 'Was anything else of importance said during this conversation?'
> A: 'Yes. He . . . mentioned Trotsky. I said, "Well, you can't put me with Trotsky." He said, "We got him, didn't we, after twenty years, and he lived in a fortress almost in Mexico," he said, "with bright lights on the place, armed guards and machine guns. But," he said, "we'll get you easy."'

Houghton was over-reaching himself and those watching him in the courtroom knew it. His evidence was increasingly greeted with smiles of incredulity and amusement. It was to Elwell 'patently false' and 'may well have been prepared in advance, possibly by Lonsdale, in case Houghton should ever be questioned by the police'.[39]

> Q: 'What was your reason for not reporting it [the meetings with the mysterious Pole]?'
> A: 'One of the reasons was that I hadn't given them – him – anything; and another was that I believed what the man told me.'
> Q: [Lord Parker intervening and studying Houghton coldly over his spectacles]: 'Do you say you are a master-at-arms?'
> A: 'Yes, sir.'

He said he ignored the next summons, another Hoover brochure, and one Friday evening in February 1958 he was going home at about 10 p.m. when he 'was accosted and attacked and beaten up badly by two men'. Despite his alleged injuries Houghton said he was too frightened to tell anyone – including Ethel Gee, 'for fear of worrying her'. Houghton went on to talk about occasional clandestine meetings over the following two years with two mysterious Russians called Nicky and John. It was Nicky, he said, who gave him the Swan Vesta matchbox with the false bottom and instructions on how to make contact if Houghton wanted to see him. Houghton asserted that he only provided them with titbits of unclassified information, missed another rendezvous and was beaten unconscious.

One of the Russians, Houghton said, warned him: '"Perhaps one day when you're having a cup of tea there will be poison in it." I forgot to mention that.' At this point, Gordon Lonsdale in the dock covered his face with his hands and his shoulders shook with suppressed laughter. Elwell undoubtedly shared his reaction. Houghton became less relaxed. He no longer stood with one hand in his pocket.

Houghton elaborated the story in his witness statement about Gordon Lonsdale coming to his house in June 1960 and introducing himself as Alex Johnson, the assistant US naval attaché, although he admitted that, later that evening, 'the penny kind of dropped. I thought, "You aren't an American."' As 7 January 1961 approached, Houghton pressed Gee to get copies of test pamphlets until he 'broke down her resistance . . . because I wanted to discuss matters with Johnson', and she provided him with seven of them the night before. Regarding the sensitive items found in Gee's handbag, Houghton denied that the test pamphlets would be of use to an enemy. He said he deliberately photographed the *Particulars of War Vessels* out of focus so they could not be read; and took the *Dreadnought* photos so 'it should be a jigsaw puzzle that no one could put together again'.

After this performance, the Attorney General – and the observing MI5 officials – must have understood that it would take little more to ensure Houghton's conviction. Houghton's defence was fatally undermined by his answers to the first clutch of cross-examination questions. 'You say you regarded yourself as a patriotic Englishman,' Manningham-Buller began. 'When did you cease so to regard yourself?' Houghton replied in January 1957, when he 'got mixed up in this'. He admitted: he did agree to provide information about the Royal Navy to people who he knew to be foreigners; he brought this information to London and supplied it on 7 January 1961; and he was guilty of agreeing to provide the information to Lonsdale, and so of the charge of conspiracy to breach the Official Secrets Act.

On the seventh and penultimate day of the trial, Lonsdale finally broke his self-imposed silence. He did not give evidence from the witness box, which would have meant him being subject to hostile cross-examination by the prosecution, but stood up in the dock to deliver a prepared statement from handwritten notes. Dressed, like Houghton, in tweeds, his stocky figure dominated the packed room as he spoke firmly in a Canadian accent but with a certain alien timbre. Elwell watched with fascination as Lonsdale accepted full responsibility for every incriminating object found in the Krogers' bungalow, even the 2,563 US dollars discovered in the attic. He said he had known the Krogers since 1955, they had become close friends, and he had given the Ronson cigarette lighter to Peter Kroger as a birthday present. While the Krogers were absent on holiday, he had borrowed their home and built the hiding place in the cellar to conceal his radio transmitter.

It was a brilliant performance. Not for the first time, Elwell admired Lonsdale's ingenuity and boldness. Lonsdale knew there was no prospect of escaping conviction himself. The Cohens faced trial under their false identities as Peter and Helen Kroger. Neither Lonsdale nor the Cohens understood how much the Security Service knew about the Krogers' genuine backgrounds. Before the jury perhaps learnt who the Krogers really were, Lonsdale had clearly thought there was a gamble worth taking: for he himself to shoulder all the blame and exonerate the Cohens. There were obvious flaws in his story, but Lonsdale had decided – no doubt after whispered exchanges with the Cohens – that there was nothing to lose.[40]

Morris Cohen spoke from a bundle of notes, acting out his assigned part of Peter Kroger, the antiquarian bookseller, with panache. He described his dealings with other booksellers and collectors, and displayed copies of his headed notepaper and the trade journal, *The Clique*, so the jury could see his advertisements. In keeping with Lonsdale's account, Cohen said his visitor had abused his trust and hospitality by filling the bungalow with spying paraphernalia.

'We did not know of secret cavities and the like,' he maintained. 'What was hidden in the books and other containers was absolutely unknown to us.' He denied all knowledge of the false Canadian passports. His style was more flamboyant than Lonsdale's, slow-paced but assertive. There was even a certain theatricality about him, occasionally waving his hands to emphasise particular points, which bordered on arrogance. As one observer reflected, he 'seemed at times to treat the court with the veiled contempt a leading West End actor might have for a provincial audience'.[41]

Lona Cohen was less impressive in her role as Helen Kroger. Perhaps, as a colourful and talented woman spy with years of experience, she instinctively found it difficult to play the part only of a dutiful housewife tending the home and assisting her husband in his business. According to her, after Lonsdale became a friend he used to help her around the house, bringing in the coal and washing the dishes, and even going shopping. In fact, she said, it was on a shopping expedition with Lonsdale on the day of the arrests that he had handed her the envelope containing the microdots discovered in her handbag. 'I know nothing of spying,' she concluded, 'and never had anything to do with such things.'

The Attorney General and counsel for the defence made their closing speeches. Behind the scenes, Manningham-Buller was fuming with anger over a bungle with the prosecution. A piece of paper had been found in one of Gee's handbags with the numbers of eighteen UDE test pamphlets relating to one particular type of sonar, and UDE had discovered that copies of all eighteen pamphlets were missing. With copies of those documents, a hostile power would have full details of that specific sonar and be capable of manufacturing it. Manningham-Buller had not been informed of the discovery before the prosecution case was over. He believed that if he had been told of this important fact earlier, he could have called evidence about it and strengthened the prosecution case even further.[42]

On the morning of Wednesday, 22 March, the Lord Chief Justice spent two hours summing up the evidence. Although his résumé was balanced, a dispassionate reader of the trial transcript decades later has little doubt of what outcome was likely. At 2.33 p.m. the jury filed out. They returned with their verdicts a little before 4 p.m. All five defendants were guilty. Before sentencing, in a carefully calculated move, Superintendent George Smith walked to the witness box to read out his carefully honed statement about the background of the defendants. It admitted he knew little of Lonsdale, but stated that he was satisfied 'he is not the man he says he is. In my opinion he is a Russian, and a member of the Russian Intelligence Service.' Lonsdale smiled broadly. When Smith went on to identify the Krogers as Morris and Lona Cohen, and outlined their links with the Rosenbergs and Abel, their disappearance and then arrival five years later in the UK with fake passports, there were looks of astonishment from the courtroom and the jury. As the superintendent spoke, some of the jurymen stared at the Cohens, their faces marked by a mixture of relief, wonder and irritation.

Only forty-five minutes after the jury delivered their verdict, the defendants stood up to be sentenced. Since the passing of the Criminal Justice Act in 1948, no prison sentence of over twenty years had been handed down. During the trial, the informed speculation was of no sentence longer than fourteen years. The Lord Chief Justice's face, however, was stony as he declared that 'for peacetime, this must be one of the most disgraceful cases to come before this court'. First Lonsdale stood to attention in the centre of the dock. The judge described him as 'the directing mind' of the five. He sentenced him to twenty-five years in prison. Some of the spectators crowded into the court gasped with surprise. Lonsdale's reaction was a faint bow and smile before he disappeared to the cells. Lord Parker described the Cohens as being professional spies and 'in this up to the hilt'. He sentenced them to twenty years each because they were older than Lonsdale. Morris Cohen's already sallow complexion became suddenly paler. Sweat was now glistening on Houghton's face and he held his head low. 'In some ways,' Parker said, his conduct was 'the most culpable', betraying 'the secrets of your own country'. The judge took into account that Houghton was 'now fifty-six– not a very young fifty-six – and it is against all our principles that a sentence should be given which might involve your dying in prison'. He was sentenced to fifteen years. Ethel Gee was last. Parker remarked that, having watched her in the witness box, he was convinced she had not acted out of 'some blind infatuation'. Rather he was inclined to think she had a stronger character than Houghton and 'acted for greed'. She too was given fifteen years.

With the verdict and sentences pronounced, the wheels of the government's publicity machine immediately began to turn. During the last three days of the trial there had in private been tensions, between the Admiralty and Ministry of Defence on the one hand and MI5 on the other, about how the publicity should be handled. MI5 objected to some of the wording in the draft Admiralty press statement, saying it was 'a good deal more definite about the compromise of information belonging to the USA and other NATO countries' than it should be, and the 'the fact is that no one knows what information may have been compromised'. With the powerful support of the Home Secretary, the Attorney General peremptorily killed off the Admiralty's pet project of a 'full-scale exhibition' featuring the actual exhibits in the trial because this might prejudice any appeal the defendants might make. Manningham-Buller agreed, however, to the release of photos of the exhibits.[43]

As the trial moved inexorably to its conclusion, with the latest revelations in the courtroom making headlines in newspapers around the world (Lord Parker himself in his summing-up described the case as 'full of intrigue' and like a 'thriller'), it became clear that the government-inspired publicity for the case and the seriousness of the issues ventilated at the trial demanded a powerful political response from the government. Such was the expected uproar, there would need to be a debate on the efficiency of the intelligence services in the House of Commons on the day after the trial. In preparation there was a meeting of the Chancellor of the Duchy of Lancaster, Charles Hill, Carrington's deputy and the Chief Whip, and intelligence chiefs (Roger Hollis and the head of the Joint Intelligence Committee, Sir Hugh Stephenson) on the last afternoon of the trial. The meeting swiftly quashed the idea of the Admiralty issuing any statement and hoped the debate in the Commons the next day could be confined to 'as narrow a front as possible', but feared there might be disclosures or allegations which necessitated the deployment of a senior minister, such as the Home Secretary. After the meeting Hill sought out Macmillan and explained 'the position fully' to him. Hill was, a senior Cabinet Office official recorded euphemistically, 'somewhat concerned to find the Prime Minister was not previously aware of all the potential difficulties of the current case'.[44]

It was hardly surprising that the Prime Minister was distracted. He had been fighting fires on four fronts at the time: crises in Rhodesia (now Zimbabwe), the Commonwealth (South Africa had withdrawn), and in Laos, 'with the possibility of war in SE Asia', as

Macmillan noted in his diary, and a collapse in sterling. Macmillan was exhausted: 'We have had 3 weeks of continuous crisis – the worst I can remember since the days before Suez . . . I have had very little sleep – sometimes not more than 2 or 3 hours a night. Once or twice . . . I have taken a mild sleeping pill.'[45] As Macmillan was to discover within days, however, his 'potential difficulties' created by Russian espionage were only beginning.

Part Four
Inquest

7

Anatomy of a spy scandal

I

Members of Parliament took their places on the benches of the House of Commons on the afternoon of 23 March 1961 in a raucous and fevered atmosphere. Day after day the media had been dominated by headlines from the spy trial, and the Labour Party Opposition scented an issue which might generate valuable political capital. MPs swiftly understood how serious a threat the government considered the Portland case posed when they saw the Prime Minister himself, Harold Macmillan, rise from his seat to make a statement and answer questions about the case. As the minister ultimately responsible for the Security Service, a consummate political actor-manager sensitive to the mood of his party, and no doubt swayed by learning the afternoon before about the 'potential difficulties' of the Portland case, Macmillan recognised that he should take the lead. The Commons fell silent. The Prime Minister began by stating that all Admiralty establishments had been directed to review immediately their security systems and that the First Lord of the Admiralty was setting up a committee of inquiry to determine who was responsible for the security lapses at UDE. He then uttered emollient words about the amount of damage wreaked by the spy ring, even though behind the scenes he had been told by MI5 that the extent of the harm caused was as yet uncharted. The Prime Minister declared that there was 'no evidence to suggest that the information compromised covered more than a relatively limited sector of the whole field of British naval weapons', no American information had been betrayed, and no secrets about nuclear weapons or nuclear submarine propulsion had been passed to the Russians. He dismissed the drawings of Britain's first nuclear submarine, HMS *Dreadnought*, found in Gee's shopping bag, as no more than plans 'of a simple electric cable layout'.[1]

Progress in the Admiralty's assessment of the damage caused by the spy ring had been snail-like. They had only started to treat the issue with any seriousness in early March, when they became perturbed by the 'public outcry' likely to follow the trial. At a first meeting to discuss this issue on 7 March it was reported that the Type 186 sonar was about 80 per cent compromised, and the Type 187 'about 27 per cent' (a remarkably exact figure in the circumstances). 'No information was forthcoming of the total scale of other information compromised.' Unimpressed, the Admiralty directed AUWE at Portland to do much better, and three days later they declared that the percentage chances of these two sonars being compromised were much higher. As for other 'AUWE projects – no compromise reported'.[2] This assessment could only have been based on speculation, because none of the spies had revealed what intelligence they stole. It reflected Portland's complacency and optimism, and the lack of understanding by many of its scientists of the strategic importance of the highly sensitive work being conducted there on the *Dreadnought*, including on her remarkable new sonar. Despite this embarrassing lack of thoroughness, the Admiralty's expert at the criminal trial only days later had no hesitation in sounding the tocsin of alarm.

The Security Service were much more prudent in assessing the likely damage. Hollis informed Macmillan that 'Miss Gee had access to classified drawings of all British asdic [sonar] sets in use and immediately projected. Houghton had access to 'Confidential' orders . . . and to classified particulars of British and Commonwealth warships. Any information of these types may have been passed by them to the Russians.' It was clear that no work on the nuclear propulsion of *Dreadnought* (based on top-secret technology passed by the Americans to the British) was taking place at Portland, so various assurances to this effect were given to important figures in the US government and military, including from the head of the Royal Navy, Admiral Caspar John, to his peer in the US navy.[3]

During the period April 1960–January 1961, while MI5 was investigating Houghton and Gee, Portland was engaged in a series of major research projects which would have been of great interest to the Soviet navy. A number of these were specifically for Britain's first nuclear submarine. *Dreadnought* was laid down in June 1959. By this date MI5 knew that Houghton (and they presumed, Gee) had already started spying for the Soviets. The submarine was launched by Queen Elizabeth II on Trafalgar Day, 21 October 1960, but it was only to make her first dive in January 1963. Urgent work at UDE on the Type 2001 Active/Passive Sonar had started in 1957 because a secret sea trial that year with the first US nuclear submarine, the *Nautilus*, demonstrated that the Royal Navy's best sonar at that time, the Type 177, could provide no protection against such a fast and deep-diving vessel. Close links with the Americans and their classified research on digital multi-beam steering ('Dimus') sonar, including work on bouncing sonar beams off the sea bed, was incorporated into the British design. This needed to be revolutionary to overcome numerous technical obstacles: for example, for the active and passive sonar to operate continuously and simultaneously over an arc of 240 degrees. (With active sonar, the system emits a pulse of sound and then the operator listens for echoes; in passive sonar, the operator listens to sounds emitted by the object he is trying to locate.) This in turn demanded massive processing and switching power. The risky decision was taken to use transistors, the new method to amplify or switch electrical signals or power that was about to revolutionise the world of electronics, rather than superannuated and bulky radio valves. The outdated Type 177 sonar used 450 electrical valves. *Dreadnought*'s 16,000 transistors were packed into twenty-eight cabinets.[4]

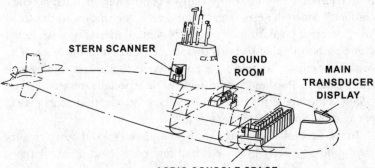

The revolutionary Type 2001 sonar from a secret UDE pamphlet of 1963. Note the enormous horseshoe of sonar in the bow and blocks of technical equipment behind.

The Type 2001 sonar was designed to meet the specific needs of *Dreadnought* and later nuclear submarines of high speed to detect enemy submarines of a similar type. Shaped like a horseshoe, the forty-foot sonar occupied the whole of the upper half of the bow of the submarine. It was to be capable of active sonar ranges at least twice those obtained for the Type 177. During the two years or so before January 1961, when Houghton and Gee were actively spying, the experimental Type 2001 equipment was subject to extensive sea trials. These achieved active sonar detection ranges of about ten miles for a submarine travelling underwater, and twenty miles for a submarine coming to the surface. They included an exercise against America's first nuclear submarine, USS *Nautilus*. For eight hours the *Nautilus* manoeuvred at 120 feet within a fourteen-mile radius of the Royal Navy test ship and during this time virtually continuous contact was held on the Type 2001 active display. The results were even more impressive for the passive sonar. In the words of one knowledgeable former submariner writing almost sixty years after the exposure of the Portland ring, the 2001 'was a ground-breaking, massively advanced sonar for its day, much admired by the Americans for the standard of its engineering and coveted by the French'.[5]

As the eminent historian of post-war Britain, Peter Hennessy, has pointed out, during the 'Cold War, the intelligence was the surrogate for hot war . . . The Cold War, in the intelligence sense, was a professionals' contest . . . a specialist's confrontation, not a people's conflict.'[6] Since the espionage front was a key theatre in the Cold War, covert information about the potential adversary was always going to be valuable to both the Soviet Union and the West. The British Admiralty should have appreciated that secret intelligence from UDE at Portland would have been especially coveted by the USSR in the late 1950s because of momentous developments taking place in the Soviet navy.

In the Second World War, the Soviet Union's naval achievements were modest, partly because it had few ports (and many of these were ice-bound in winter) despite its gargantuan coastline. In 1945 Stalin started a programme of naval expansion, with a major focus on submarines. He ordered hundreds to be built, based on the latest German designs captured after the war. Only a limited number of the submarine orders were completed by the time of Stalin's death in

1953, when the naval building programme for big ships was halted. With Khrushchev's rise to power in 1956 it was jolted back into life. The new Soviet leader was a strong enthusiast for submarines, writing in his memoirs: 'A submarine . . . looks like a floating cigar. But the submarine is the supreme naval weapon nowadays, and I'm proud of . . . introducing submarines as the basis of our sea power.' To spearhead the growth of Soviet naval, and particularly submarine, power he appointed Admiral Sergey Gorshkov commander-in-chief of the Soviet navy in January 1956, at the relatively youthful age of forty-five. Within a year or two it was clear to the West that Gorshkov was initiating a programme to modernise and enlarge the entire Soviet fleet (both submarine and surface). It was in 1958 that the first Soviet nuclear submarine, the *K-3 Leninsky Komsomol*, made its maiden voyage, the prototype of the Project 627 November class.[7] These factors all help to explain why underwater maritime espionage against the Western powers was a priority for the KGB: the flow of sensitive information from Houghton and Gee could not have been more timely.

In early March 1961, before the Portland spy trial and with his first meeting with President Kennedy only weeks away, Macmillan was perturbed by the effect of the Portland spy case on security relations with the Americans and asked for a Security Service briefing in preparation for his talks with Kennedy. Hollis was requested 'to take the offensive by calling attention to our successes and, in particular, perhaps to the fact that you have caught people who have slipped out of the hands of the Americans themselves'. The tone of the MI5 brief was diplomatic, highlighting the counter-espionage successes and failures of both the British and American services, summarising the Portland case, and saying MI5 would welcome new security discussions with the Americans, but only on the understanding that they 'should cover American security procedures as well as our own'.[8]

Macmillan recognised after the debate in the Commons on 23 March, when MI5 and the Foreign Office were castigated, that the Portland case required a weighty political response. He immediately advised Lord Carrington as First Lord of the Admiralty not to travel on a planned official visit abroad before 'the problem of the tribunal was settled'. The next day Macmillan flew out to the US naval base at Key West in Florida for his first meeting with Kennedy, rushed forward at the president's request because of the political crisis in Laos. The talks lasted about three hours. Although their discussions focused on Laos, Macmillan's private diary confirms they conferred over various other 'things'. These almost certainly included the Portland case and its security implications, because the Prime Minister had ordered the special Security Service briefing in response to David Ormsby-Gore's warning from Washington only a few weeks before.[9]

After discussion in Cabinet, responsibility for the inquest into what went wrong in the Portland case was passed to the Home Secretary, Rab Butler, as the minister responsible for national security. Butler selected a chairman for the tribunal the following week (Sir Charles Romer, a retired Court of Appeal judge). On 29 March he announced its creation and that it would report to the Prime Minister. The inquiry was to investigate the breaches of security at Portland and who was responsible for them, any breaches in existing security procedures, and how and why Gordon Lonsdale and the Cohens were allowed into the UK.[10]

The day before the inquiry was announced, Houghton asked his barrister to pass to 'the proper authorities' a note which was to haunt the Security Service in future years in unexpected ways. Houghton listed intelligence which he said the Russians required (including on homing torpedoes, a listening post in the Straits of Gibraltar, devices towed behind vessels to disguise engine and other noises, a chain of sonar buoys from Greenland/Iceland to Cape Wrath, and underwater sonar); and information which the Russians were already receiving (such as on sea mines, radar, the British nuclear submarine *Dreadnought* 'in UK and USA', and the activities of American Polaris missile submarines at Holy Loch in Scotland – where the US navy had a ballistic missile refit base from 1961 to 1992). Against the heading 'Intelligence the Russians were already receiving', Houghton had commented: 'Possibly by the three persons I could identify.' Houghton asked his barrister to add to the memo the address in Lancaster Road in Notting Hill where the KGB officer Dozhdalev had lived. The note was sent to MI5. Houghton was clearly manoeuvring in an attempt to reduce his fifteen-year sentence by offering to identify some KGB officers he had dealt with. D1 immediately passed this information to the Admiralty, stressing that MI5 had not yet interrogated Houghton and therefore could not 'comment or elaborate on what he has written'. The Security Service could not interview Houghton about the note immediately for fear it would jeopardise the prisoner's legal appeal against his sentence. This delay, however, was to prove only the first problem for MI5 created by Houghton's memo.[11]

Within days rooms were commandeered in the Cabinet Offices in the heart of Whitehall for the Romer Committee to take evidence, a schedule of witnesses was drawn up, and it was decided the inquiry should work on an administrative rather than judicial basis (meaning that none of the witnesses would need to be represented by lawyers).[12] In the coming weeks, over twenty witnesses were summoned to give evidence of extreme sensitivity, including Roger Hollis (no fewer than five times) and Arthur Bonsall of GCHQ. All their testimony, marked 'Top Secret' at the time, was recorded by a stenographer. The documents which resulted form a remarkable historical record of a modern democracy holding an inquest into the lessons to be learnt from successful penetration by a hostile intelligence service and from failures of its security procedures, and of its willingness to make that record public.

Elwell drafted a lengthy document, 'The Lonsdale Case: A Preliminary Report', which MI5 provided to the inquiry.[13] It encapsulated the Security Service's knowledge about the spy ring at the end of March 1961. Concerning the possibility of other unidentified agents, the report was clear: 'there is no evidence that the network included others during the period when it was under investigation'. It included an updated biography of Lonsdale, reflecting the analysis of the letters to and from Lonsdale contained in the microdots and manuscript found in Helen Kroger's handbag. Four of these letters were missives to Lonsdale from Russia: one from his mother; two letters from Lonsdale's wife, Galisha, one dated November and one December 1960; and one from his daughter, Liza. The six-page manuscript was Lonsdale's response to those letters. They were in fluent Russian and the Attorney General had read out some translated extracts at the start of the trial. The first important clue was that 'Lonsdale' had signed off the correspondence to his wife with a 'K', clearly the first letter of his Christian name, and not G. The letters from Liza and Lonsdale's mother were not read out in court. Liza's letter throbbed with yearning, both hers and her evidently younger brother, Trofim's, for their father to return to Russia soon. Trofim added a heartfelt postscript to his sister's letter: 'PS. Daddy, come home quickly.' As a devoted father with four young children of his own, reading such passages must have moved Elwell and enhanced his sympathy for the Russian agent, exiled from his homeland and his family.

None of the correspondence contained an address or any family names. They were crammed with gossip – the health of relations, the children's progress at school, outings – but also understandable complaints about their enforced separation and pleas for Lonsdale to bring home for his wife some luxury goods from the West, like 'a white brocade dress, with white shoes'. Also in the letters were potential clues to Lonsdale's identity and peregrinations before his arrival in the UK in early 1955. In her first letter Galisha wrote that 'Mother' was sick, and Lonsdale 'must somehow make haste to come home – one has only one mother in this world'. His mother was clearly still alive, living in Moscow, and apparently a widow. Galisha described her work at a school and how she had recently sung at a concert there, which 'reminded me of our life in Prague, and I felt very sad . . . If my memory does not deceive me it is already seven Octobers and six May Day celebrations, and this does not include family festivities, that I am alone.'

The references to the missed May Day and October Revolution celebrations helped narrow down the time when Lonsdale embarked on his mission to Britain. Elwell assumed Lonsdale must have left Russia between 1953 and 1954, before spending five or six months in Canada in Vancouver and Toronto to obtain his fake passport, crossing the Canadian border by train at Niagara Falls into the USA in February 1955 and then travelling on to New York and Britain shortly afterwards.[14]

Lonsdale's reply to his wife provided Elwell with other clues: 'From the age of ten, during the past twenty-nine years, I have spent only ten years with my own people. I did not wish it and I did not seek it, but so it turned out to be . . . I have thought very much about it – why all this? The answer is it all started as far back as 1932 when Mother decided to despatch me to 'Tartarare' [translated as 'the nether regions']. At that time she could not imagine, of course, all the consequences of this step and I do not blame her.'

He continued to describe his cramped service flat in the White House before commenting that it was 'the eighth year I have celebrated without you. Such is life ... [signed]"K"'. The letter ended with a significant postscript, 'I will be thirty-nine shortly. Is there much left?' The 'nether regions' Lonsdale referred to was a Russian idiom for a faraway place. Elwell was convinced it meant that Lonsdale had been sent to a country long distant from home when he was a boy, possibly America or Canada. This would explain his almost flawless English and why the KGB wished to recruit him. By stating that he would be thirty-nine shortly, the 'Lonsdale' in prison revealed he was born in January 1922, not August 1924 like the genuine Gordon Lonsdale whose identity he had purloined.[15]

Other information confirmed that Lonsdale had met the Krogers in Paris in April 1955, before the Krogers left on a lengthy trip to Hong Kong and Tokyo. Although his direct dealings with Lonsdale so far had been very limited, Elwell did not hesitate to sketch out his sympathetic view of the Russian illegal: 'A man of considerable charm, Lonsdale won the friendship of a large number of men and women. In his conversation he is humorous and ironical rather than witty, ready to talk fluently and trenchantly on most subjects ... as a businessman he won the respect and confidence of his frequently shady colleagues.'

Elwell's character sketch of the Krogers, on the other hand, was excoriating, again based on little or no personal contact, but undoubtedly stemming from his abiding dislike of Westerners who had been seduced by Soviet communism. He painted Peter Kroger as 'a sententious bore, eaten up with pedagogy, a scribbler of sloppy verse, a man whose life appears to be governed by rancid idealism ... Mrs Kroger is even less alluring ... she looks and probably behaves like an embittered crazy fanatic.'

GCHQ pored over the Krogers' wireless equipment, and the cipher pads and signal plans discovered at 45 Cranley Drive and Lonsdale's apartment. Some of the signal plans were for transmissions in Morse from Moscow and from the illegals, and for two-way transmissions between the agent and the Centre. GCHQ intercepted one of these from Moscow on 15 January 1961 and discovered that – unlike others sent after the arrests – it contained traffic. Although it is probable Cheltenham deciphered this message, whether it in fact did so is still a classified secret. GCHQ also made various detailed observations and tests on the transmission equipment: it was 'a highly efficient, beautifully engineered system capable of worldwide communication at any time of day or night'.[16]

II

Reading the transcripts of evidence to the Romer Inquiry decades later, one detects the dull but unmistakable thud of bucks being passed and the understandable attempts at self-exculpation typical of all bureaucracies – including supposedly secret ones – defending their conduct, and themselves, against criticism by forces beyond their control. Potential targets in the cross-hairs were MI5 (for not detecting the spies earlier), GCHQ (for not intercepting traffic from the Cohens' undercover radio station in Ruislip), the Passport Office (for permitting Lonsdale and the Cohens to travel to and from the UK undetected on false passports), but principally the Admiralty (for appointing Houghton to UDE in the first place, and later allowing him and Gee to carry on their espionage at Portland for years).

The parade of Admiralty witnesses on the whole told a sorry tale. The Admiralty had lost the crucial annual report about Houghton's conduct while he was in Warsaw, and its appointment of Houghton to UDE was (in the words of the committee chairman) 'a slap happy affair', illustrated by the fact that none of the correspondence sent to London from Warsaw in 1952 documenting Houghton's drunkenness and weak performance was brought to the attention of either the Admiralty staff appointing Houghton to Portland or to MI5 when they did a cursory check on him in 1956. The Admiralty's application of its security procedures was amateurish (with no central coordination point for security, and no system of inspection of security arrangements at locations like UDE from headquarters). The Director of Naval Intelligence, Rear-Admiral Nigel Denning, frankly admitted his impotence and confusion: he found it 'very difficult, owing to the divided responsibilities in the civilian sections [of the Admiralty], to know exactly who is responsible for what'. Witnesses from Portland attested to the lack of rigour and interest at Portland in following up Mrs Houghton's allegations, made both before and after the incident of the 'missing files' in 1956, that her husband was passing secret UDE information to communist agents in London, or other indications that Houghton was a security risk:

they did little or nothing at the time, assuming it was the responsibility of the Admiralty in London and MI5.[17]

Perhaps for this reason, the inquiry – rightly – treated Houghton's ex-wife as a witness with extreme courtesy. In further clues to Houghton's motivation for his treachery and his character, she remembered him once in Warsaw uttering, 'I would do anything, work for whichever side paid the most money,' and his abiding bitterness and resentment at being reprimanded and sent back to the UK. She gave evidence, which clearly troubled the tribunal, about how her attempts to have her concerns heard and investigated by Houghton's manager and the local probation officer were rebuffed.[18]

Great deference was paid to Arthur Bonsall of GCHQ, who was asked by the committee's chairman to explain 'his mysteries'. Bonsall did so patiently. No great technical skill, he confirmed, was required by the Krogers to operate the radio equipment found buried in their cellar. He described the one attempt made by GCHQ in Ruislip in December 1960 to detect radio transmissions and why it ended in failure, and even demonstrated the 'squash transmission' technology used by the KGB to an awed committee. The Russians had tightened up their radio procedures around 1959, Bonsall revealed, so that 'in general their awareness of the need for the greatest possible precautions in the use of radio, and the need for complete security, is outstanding'. They transmitted messages 'into the blue' to their agents all around the world at fixed but unpredictable times and the agents never acknowledged them. KGB agents in turn sent back any messages to Moscow at set but unpredictable times in accordance with their signal plan. Neither party needed to establish contact before transmitting. Wavelengths used by Russian agents were particularly inconvenient for GCHQ when used in the UK. To detect them one needed to be either fairly close, within around fifteen miles, or about 400 miles away, when the signal bounced down again – which normally meant beyond the shores of the British Isles.[19]

The Security Service Hollis described to the inquiry in his introductory bout of evidence was a completely different organisation to the MI5 of the 2020s: it was much smaller, with officers recruited to serve their 'whole life in the Security Service', and no contemporary DG would begin by stressing the extent to which, although his minister was the Home Secretary, his organisation enjoyed 'glorious independence'. Counter-espionage, he explained, at its simplest consisted of catching spies (over which MI5 had 'almost a monopoly') and protecting secrets, which was a duty shared with government departments, who were primarily responsible for their own security. When asked directly if MI5 had enough staff, Hollis (who only had 174 officer-level staff in total, and was instinctively cautious) replied, 'very nearly sufficient'. Combating Soviet espionage was a key responsibility of the Security Service and Hollis was clearly proud to tell the inquiry that the head of his Soviet section, D1, Arthur Martin, had been despatched to Washington at the invitation of Hoover to brief the FBI (and CIA) on MI5's knowledge of Russian intelligence, which he said was 'very much in advance of their [the FBI's] central study section, and they want to set up something' similar.[20]

Hollis condemned the slack security practices at Portland, such as allowing senior staff to take secret documents home every night to work on them. A major focus of one session with Hollis was the Admiralty's request in 1956 to vet Houghton and the Security Service's response that his ex-wife's allegations seemed 'mainly spiteful'. Hollis calmly defended MI5's action at the time, arguing that the Security Service interpreted the notes from the Admiralty as making 'a rather vague allegation', requesting MI5 only to carry out a check on Houghton in the Registry, and confirming that Portland was keeping the clerk under 'discreet surveillance'. He admitted, however – with classic understatement – that MI5 'had no reason to express a view one way or the other' about Mrs Houghton's motives and that the comment about spitefulness was not 'particularly sensible'.

Hollis's view was that in 1956 the problem lay with omissions of reporting and central recording, failures to collate the various incidents and concerns about Houghton as they happened in one place, and assess them 'as a whole'. If this had happened, the investigation into Houghton might have started much earlier in 1956. He also thought the preliminary enquiries the Security Service had started about Houghton in early 1960 – independent of the precious intelligence from Goleniewski – would have led them to put the clerk under surveillance and guided them to Lonsdale.[21]

Hollis stoutly protected the Security Service against press censure for failing to identify the Cohens earlier after the FBI had sent their fingerprints to the UK in 1958. He pointed out that this would have been impossible because the Cohens had travelled to and from the country as the Krogers on genuine New Zealand passports, and they were only recognised as the Cohens on 11 January 1961 after their arrest and their fingerprints had been taken. To the obvious fascination of the committee, he showed them examples of microdots and how they could be concealed beneath postage stamps or even built into paper. Hollis said he was 'pretty confident' that, with the arrest of Houghton and Gee, the 'whole of Lonsdale's agent network' had been rolled up. Another theme of Hollis's was that, to be successful, counter-espionage 'needs luck'. The metaphor was, typically for a man of his generation and background, taken from cricket: 'not pure luck. Obviously a well-organised security organisation is very ready to hold its catches when they come even if they are pretty quick ones . . . But it is almost impossible to guarantee to catch a spy,' especially when a trusted employee 'goes bad on you'.[22]

Hollis was the last person to give evidence to the Romer Inquiry, on 25 May. In the two months since it had been set up, the political turbulence created for the government by the Portland spy ring had become a storm. The cause once again was espionage. Late on the afternoon of 5 April, in the greatest secrecy, MI6 officer George Blake had confessed to being a Soviet spy for the previous eight years, regularly passing information about the activities and organisation of MI6 and the CIA to the Russians. In the weeks that followed, MI6 research concluded that Blake's espionage had cost at least forty lives. Such momentous treachery by a long-serving and popular MI6 officer, when made public so soon after the Portland trial, was certain to create a political tempest. The Prime Minister was in America on his official visit to President Kennedy at the time. Since 1956, when, as Foreign Secretary, he had been reluctantly forced to make a statement in the Commons clearing Kim Philby of being a Soviet agent, Macmillan had an instinctive dislike of spy scandals. He distrusted (and was later actively to hate) the brouhaha they whipped up in the media and the embarrassment created for the government of the day, and considered it was bad for the national interest for spies and their treasonous activities to be debated in public. 'The government could fall on this,' was Macmillan's first reaction when told of Blake's confession by the head of MI6. His growing annoyance at espionage scandals was sharpened by the need personally to inform the president. The story as told by MI6 veterans is that, when asked to break the news about Blake to Kennedy in a discreet way at a meeting in Washington in early April, the Prime Minister murmured gnomically that '"C" [meaning the head of MI6] has nabbed a wrong 'un.' Kennedy was understandably flummoxed in response.[23]

After Blake's exposure, Macmillan exploited the British success in arresting the Portland ring as a way of reassuring the Americans about British security. In the wake of the conviction of the three KGB illegals (two of them American), even J. Edgar Hoover was surprisingly understanding. The Blake case, he told the Security Service, was a 'further illustration of how constantly alert we had to be to the dangers which beset us. After all . . . Christ Himself found a traitor in His small team of twelve . . .'. Not all parts of American government were reassured, however. Soon after the Portland trial, at a meeting in Washington of the National Security Council the US navy made an attempt to force through a significant break in the exchange of UK–USA intelligence. It was defeated partly because of intervention by Alan Belmont, among others.[24]

Macmillan understood that other important audiences needed reassurance about security: the British Parliament and voters. The Romer Inquiry was already taking evidence by the date, 3 May 1961, when Blake was convicted at the Old Bailey after a largely secret trial and sentenced to forty-two years in prison. Eight days later – after initially opposing any further enquiry, but then taking soundings with the Home Secretary and the Labour Opposition – Macmillan bowed to the inevitable: he announced the setting up of a committee under the eminent lawyer, Lord Radcliffe, to carry out a comprehensive review of protective security across the whole of the public services. Its other (unpublicised) task was to scrutinise the facts of the Blake case, and to parallel the work of the Romer Inquiry into Houghton and Gee. On 30 May the Romer Committee summarised its main findings for the Cabinet Secretary so he could advise the Prime Minister. The Romer Report:

(i) Finds there was a general lack of security-mindedness at Portland in 1956 and criticises the Captain in charge at the time, both for this and for certain particular omissions;
(ii) Severely blames the Security Officer and one other member of staff at Portland for their conduct;

(iii) Blames the Admiralty both for the manner in which they exercised their responsibilities for security at Portland and also for the inefficient and casual way in which security matters are handled in the Admiralty;

(iv) Exonerates the Security Service (apart from one comparatively minor criticism relating to 1956) and commends their skill in catching Lonsdale and the Krogers;

(v) Exonerates the Immigration authorities;

(vi) Exonerates GCHQ;

(vii) Finds that there was no failure of liaison with the United States or any other country on security matters.[25]

A copy of the report was passed immediately to the Radcliffe Committee. Otherwise, circulation of the numbered copies of Romer was tightly restricted to a clutch of senior ministers and the heads of MI5, MI6 and GCHQ. The main target of the inquiry's stinging criticism was the Admiralty and some of its staff at Portland. Chief among these was Captain Pollock, head of UDE in 1956. The committee held him responsible for the general want of 'security-mindedness' at UDE, and the security officer at UDE was 'gravely to blame' because he never interviewed Mrs Houghton or the regional welfare officer to whom she had talked, and, without any personal investigation at all, drafted the letter to the Admiralty suggesting that Mrs Houghton's suspicions might be the 'outpourings of a jealous and disgruntled wife'. With the exception of the female local officer who had been dismissed as an untrustworthy alcoholic by her colleagues – all male – when the Security Service started their investigations at Portland in 1960, 'everyone to whom Mrs Houghton communicated her strong suspicions . . . attached no importance to her story'. The committee emphasised that they were 'impressed' by the testimony of Houghton's former wife and, in a gentle swipe at MI5, found the Security Service had no grounds at all to suggest that her allegations seemed 'mainly spiteful'.

To the contemporary eye, the original dismissal of the allegations of Houghton's ex-wife and the motivation ascribed to her read like egregious sexism, but of course they would not have appeared so to many in the 1950s and early 1960s. Marriage was the norm (in 1961, out of every thousand women aged 21–39, 808 were married, compared interestingly to 572 back in 1931).

Courtship and marriage, followed by a life at home looking after children, were the main aspirations for the overwhelming majority of women, conditioned by the beliefs and expectations of a society which was still unequivocally male. When the Cambridge-educated novelist Jessica Mann was about to have a caesarean in 1961 it was not she, but her husband, who was requested to sign the consent form. In the same year, Lord Ailwyn declared in Parliament, 'If I had my way, I would introduce a law forbidding mothers with young children to leave them all day and go out to work.' More women than ever, in fact, were in the labour force (thirty-three per cent in 1961; two per cent more than a decade before), but many were in part-time jobs and placed in lower grades with lower pay than men. In 1961, a handful of women were starting to storm bastions of male privilege in the professions, journalism and the establishment, but they were as noticeable for their rarity as for their outstanding achievements.[26]

Although some remarkable women worked in MI5, a meagre handful were of officer rank and the prevailing misogyny blocked any being promoted to more senior positions until the 1970s. In the early 1960s, some of the more enlightened men in the Security Service encouraged capable female secretaries to spread their professional wings (one in counter-espionage, for example, had regular meetings with a former Polish army officer who was a Security Service agent) but they were few. The prevailing orthodoxy in MI5 and its conflict with the growing assertiveness of young women were illustrated by Alex Kellar, Director of 'Communism – home', complaining to Personnel when his fashion-conscious new secretary came to work wearing a black leather skirt (which covered her knees) and red socks. When chided by Personnel, rather than acquiesce, twenty-year-old Caroline Macmillan resigned on the spot.[27] For young women, the post-war world 'promised much while delivering little', as social historian Victoria Nicholson has pointed out. 'Sex was taboo outside marriage. Unwed mothers were shunned, contraception was reserved for wives and pre-marital intercourse was only for sluts.'[28]

The Romer report enumerated the many security failings at Portland, and the Admiralty overall was lambasted for its lackadaisical approach to security and for having no centralised point of coordination for this important responsibility.[29]

Romer lavished praise on the 'close and effective international security liaison' between MI5, the USA, Canada and other countries – the detection of these KGB spies being a 'convincing example' of how it worked in practice – and on MI5. The Security Service had not overlooked any 'lead' which could have led it to identify Lonsdale or the Krogers earlier, and showed 'professional skill of a high order' in establishing the necessary evidence against them.[30]

Having read the Romer Report, Macmillan's primary concern was not the Security Service. The report had vindicated his view at the time that MI5 was an effective organisation. When briefing Opposition leaders on espionage matters around the time the Romer Committee was being set up, Macmillan had told them that he did not think that an investigating committee would 'need to probe into the efficiency of MI5 operations. It seems probable that they will find out that MI5's detection work was, in this case, competently carried out. Any need for investigation arises on the other side of the picture, in security procedures and their application in the Admiralty and its out-stations here and abroad.' The Prime Minister's prediction proved uncannily accurate, and as a result he needed to salvage the political career of the First Lord of the Admiralty. Lord Carrington was a rising politician (he celebrated his forty-second birthday in June 1961) whom Macmillan liked and respected. Ever the honourable man, Carrington submitted his resignation, but (unlike in 1982 after Argentina invaded the Falklands, when he was Foreign Secretary and his resignation was accepted) the Prime Minister refused it. Macmillan 'wouldn't hear of it', Carrington later recalled. The Prime Minister decided his best strategy to cauterise the political wound was, on 13 June, to publish only a very brief summary of the report's conclusions; on the same day in a statement to the Commons, to shoulder ministerial responsibility for the security scandal together with the Home Secretary and Carrington; and to brazen out the inevitable attacks from the Labour Opposition and embarrassing questions from backbench MPs. The tactic succeeded brilliantly. Only the talented but often drunk deputy leader of the Labour Party, George Brown, came close to landing a blow that day.[31]

Hollis must have purred with a combination of satisfaction and relief when he first turned the pages of the Romer Report. He had appeared five times before the Romer Committee and, in his unassuming way, given evidence confidently and with an obvious command of the facts. Under his direction the Security Service had not only avoided any painful brickbats but garnered high praise. He understood, however, that MI5 needed to respond quickly to what muted criticism of the Service there was in the report. He sent an office circular around the whole organisation in mid-June underlining the praise of the professionalism of the Security Service in the Romer Report, but – in response to the report's chiding of MI5 for its response to the Admiralty in 1956 about Houghton's ex-wife – issued a 'general warning to officers against expressing opinions in official correspondence when they have inadequate evidence . . . on which to base such opinions'. Referring back to the Security Service's response to the anti-Semitic letter sent to the worker at Portland in January 1960, he also instructed MI5's counter-espionage branch 'to take further or to look into all reports and allegations within their province, even though they may be vague or nebulous'. Hollis ensured the Prime Minister was informed of these steps.[32]

MPs were able to question Macmillan in Parliament a few days later. The Prime Minister was again at his suave best, accentuating the various security reforms in train at the Admiralty, how all defence departments had reviewed their security, and defending the Security Service as 'on the whole . . . remarkably efficient'. Armed with the report summary, the Labour leader, Hugh Gaitskell, harried Macmillan over his decision not to accept Lord Carrington's resignation. The Prime Minister stood firm. Carrington survived, and in the coming months oversaw the belated introduction of a centralised Security Department in the Admiralty and the appointment of a new Director of Security.[33]

The disciplinary hearings against the Admiralty employees censured in the Romer Report trundled on throughout 1961, but with unexpected outcomes. Although the Portland security officer was dismissed from his post, as a temporary civil servant his pension was not cut; and the head of UDE in 1956, Captain Pollock, who retired in 1958, submitted a robust defence. Almost a year after the Portland trial, the Admiralty decided there were simply no grounds for disciplinary action against him.[34]

The lessons of the Portland spy ring were not studied in the UK alone. NATO, for example, had a Special Committee during the early 1960s which met every six months to debate and exchange information about espionage, subversion and sabotage by the Communist Bloc. The counter-espionage services of NATO members, including the FBI, MI5 and RCMP, sent delegates. When the committee met in Paris at the end of May 1961, it discussed the Lonsdale and Blake espionage cases and agreed on a new study of 'The Use of False or Improperly Obtained Personal Documentation by the Soviet Bloc Intelligence Services and Communist Parties'. All NATO members were to submit papers on the subject. Within Canada itself Lonsdale's fake passport prompted a new round of soul-searching, and the RCMP made a series of radical suggestions to the Canadian Cabinet in response, including fingerprinting all Canadian citizens at home and abroad. All were rejected.

The NATO Special Committee quickly decided, when it met again in November, that there were 'considerable difficulties' in the way of substantially tightening up passport security. The committee had to content itself with NATO member countries swapping information about the latest Soviet Bloc methods to obtain fake passports, and vague recommendations that secret marks in genuine passports and fingerprinting should be used more widely.[35]

With the completion of the report of the Radcliffe Committee, any further security lessons to be learnt from the Portland spy case were spliced into its conclusions. These focused on more protective security in all sensitive government jobs, and in particular more positive vetting. Hollis was sent his copy of the Radcliffe Report in December 1961 and he must have sighed with relief that the Security Service had, again, emerged free of criticism from an official inquiry. His only concern was whether the committee's recommendations would lead to MI5 being inundated with more protective security work: although a quiet and reserved man, Hollis knew his way through the Whitehall thicket and had started lobbying for more funds from the summer of 1961. His efforts were rewarded. Largely as a consequence of the counter-espionage cases of 1961–2, the Cabinet Secretary was to approve the Service recruiting an extra fifty officers, and 250 additional staff in other ranks and grades.[36]

Of course, Western intelligence services knew it was highly likely that they, and their political and military masters, were not alone in holding an inquest on the arrest of the Portland spy ring and learning lessons from it. Behind the Iron Curtain, they would have been sure that the KGB and the Kremlin were mirroring these exercises.

The difference, of course, was that in Moscow such inquiries would take place behind a hermetic shield of secrecy, and any conclusions take years to be revealed, if they were ever to emerge at all. While the high politics of the statecraft of espionage were proceeding throughout 1961, so were the low politics and cunning of spying itself: unmasking the real identity of Gordon Lonsdale, and squeezing any of the five convicted spies into a betrayal of their work, their secrets and their Russian masters in the KGB.

Part Five
Prison and Spy Swaps

8

Gordon Lonsdale

I

Knowledge of Lonsdale's true identity would be an invaluable weapon in the duel of wits and resilience which Charles Elwell hoped to play out with him in the coming months and perhaps years, to persuade the Russian agent to cooperate: hopefully to reveal sensitive information, perhaps to defect, perhaps even to send him back to the Soviet Union as a double agent. There were, in Hollis's words, 'an immense number of things' the Security Service wished to learn from Lonsdale and the Krogers.[1]

MI5 could not make a premature approach because the five prisoners appealed against their sentences. All the appeals were finally rejected on 8 May 1961, the sole query of the judges being whether Houghton's punishment was too lenient.[2]

Incarcerated first at Brixton prison south of the River Thames after his arrest, then at Wormwood Scrubs in west London, Lonsdale was driven up to Strangeways prison in Manchester in June. He was later transferred to Birmingham. Around the same time the Krogers and Gee were moved to prisons in Birmingham, while Houghton was transferred to Winchester. During their early days in prison – despite representations by MI5 soon after the trial ended – the British authorities were remarkably lax in their treatment of imprisoned KGB spies. An MI6 officer who visited George Blake in Wormwood Scrubs towards the end of May reported to the Security Service that Blake had told him 'he was able to have a chat with Lonsdale during exercise breaks and occasionally in the canteen. Blake said that Lonsdale seemed quite a pleasant fellow ... and mentioned that [Lonsdale] considers his downfall was due entirely to the ineptitude of Houghton.'[3] George Blake later recalled his first meeting with Lonsdale in the exercise yard, when he was trudging round 'in the centre of a vast, slow-moving circle of some seven hundred chattering, shouting and shuffling prisoners'. Lonsdale came across straight away to shake hands and introduce himself – a 'stockily built man of medium height with a broad, cheerful face and very intelligent eyes' – and they discussed the chances of being released early. For Lonsdale, Blake wrote, 'there was always a chance that . . . an exchange might be arranged'.[4]

From studying intercepted phone calls and correspondence over the months, Elwell knew that the person who visited Lonsdale in prison regularly on Saturdays, and for whom he appeared to have most respect, was his former business partner, Molly Baker. She agreed to act as an intermediary to gain Lonsdale's trust. Lonsdale confirmed to her that he was prepared to provide sensitive information in exchange for a 'substantial reduction' in his and the Krogers' sentences. He was 'not prepared to bargain for himself alone', nor 'to give information which would lead to the arrest of other persons'. Aware that gossip was rife in prison, Lonsdale stressed that the interviews should be conducted in the greatest secrecy.[5]

MI5 held delicate discussions with the government on what offer could be made to Lonsdale to persuade him to divulge valuable intelligence. In early May Elwell drafted a paper which reflected his buccaneering and imaginative spirit. He proposed a tariff of reductions in Lonsdale's severe twenty-five-year sentence (Elwell was sympathetic to the Russian's position because of his own lengthy imprisonment during the Second World War in Colditz) calibrated to the level of assistance given (for example, five years for identifying an active spy against the UK). But, much less conventionally, as a way of reducing Lonsdale's sentence, Elwell suggested that the government could 'allow him to escape'. Needless to say, MI5's legal adviser scotched this idea as unlawful. When Hollis finally met the top civil servant at the Home Office, Sir Charles Cunningham, he was unbending. It would be 'quite wrong for us to strike any kind of bargain', he emphasised, but if the spies gave MI5 valuable information, 'this should be taken into consideration' by the Home Secretary when their sentences were reviewed.[6]

The first interview was arranged for 31 May 1961 at Wormwood Scrubs. Elwell (known as 'Mr Elton of the Security Service') and Molly Baker met Lonsdale in a room with brown paper stuck over the glass in the door and the windows in a well-intentioned if feeble attempt to maintain security. Wary, Lonsdale insisted that, before

Molly Baker left the room, she repeated the burden of his message to MI5, and in particular that the Krogers must be included in any deal. Elwell, having deposited his bowler hat and umbrella in the deputy governor's office but wearing his trademark dark suit, was finally left alone with Lonsdale in his prison fatigues. He offered the prisoner some cherries from a bag he had bought on the way and a copy of a new book about the spy ring.[7]

Elwell began by showing the KGB illegal the photograph Lonsdale had taken of him at his rented flat in July 1956, which had caused Elwell short-lived embarrassment when discovered in Lonsdale's bank deposit box. Lonsdale recognised it as having been taken in Elwell's flat and commented, 'You will then be on our files in Moscow,' explaining that he had sent back to Russia all his photos and that the KGB had instructed him specifically to enrol at SOAS 'to get to know members of the British Intelligence Services who were studying there'. Lonsdale outlined his supposed career in the KGB's illegal branch, suggesting he had joined in 1940, operated against the Germans and after the war served in China. He offered to divulge intelligence on the careers and activities of the Portland spy ring, the organisation and methods of Soviet illegals and counter-intelligence, and, tantalisingly, about 'another agent whom he had run up to the time of his arrest', the location of his second wireless transmitter and an 'auxiliary illegal network with which Lonsdale had worked'. When Elwell revealed how the British were unwilling to offer anything certain in return, and how nothing Lonsdale could reveal would help the Krogers, Lonsdale's reaction was 'frankly derisive'. The spy said he was very apprehensive about the Russians learning that he was having conversations with a member of MI5, and stressed the serious peril he faced of 'being shot when he returned to Russia' if he was freed early under some arrangement with the British government.

He went on to suggest that he suspected he had been under surveillance from the time his belongings in the bank were searched, and had reported his suspicions to Moscow, but had been instructed to remain in the UK. The prisoner painted a picture of the Krogers as being minor and fairly inept agents, but his particular scorn was reserved for Houghton: 'the worst man he had had dealings with during his career as an intelligence officer . . . though cunning he was stupid and completely self-centred', and most of his evidence

in court was 'pure fabrication'. Lonsdale said the Admiralty clerk
had provided to the Russians around 350 test pamphlets about
anti-submarine equipment, 'including some connected with British
atomic submarines', and that the KGB Illegals Directorate had
about 300 officers like himself, of whom 'a fair proportion' would
be operating in Britain but the 'vast majority' in the USA. The meeting ended on a gloomy note: Lonsdale saw little hope of coming to
'any satisfactory arrangement'.

A fortnight later, Elwell met the Krogers at Winson Green prison
in Birmingham. He started by asking them if they thought they had
been fairly treated. Peter Kroger said he considered their sentences
excessive, to which Elwell – known for his directness – asked if
he thought 'they would have got off any more lightly if they had
been caught in similar circumstances in Russia'. Helen Kroger
complained that their possession of New Zealand passports should
not have been held against them because they were genuine. Elwell
retorted that, although the passports were valid, her birth certificate
was not. At this point Peter Kroger 'intervened hastily and told her
not to discuss the matter'. To every query, and whenever any tension
developed, Kroger repeated, 'Go and see Lonsdale.' Elwell found the
Krogers' attitude 'made conversation rather heavy going and [he]
withdrew from the fray for a time to allow the Krogers to talk to one
another while [he] prepared to return to the charge'. Elwell decided
that the Krogers knew of Lonsdale's plan to try to barter with the
British authorities and was not therefore surprised when Peter Kroger asked for a meeting with the Russian spy. This time Elwell found
them 'both polite and even pleasant. Mrs Kroger is a more agreeable
person than Kroger, whose manner I found both stubborn and sly.
He looked rather old and unwholesome. He would be a very red rag
to an anti-Semitic bull.'[8]

Elwell's next meetings with Lonsdale took place at Strangeways
prison in Manchester in late June. He spent seven hours with him
in total. The MI5 officer referred to information that Lonsdale had
helped the Krogers choose their bungalow. The prisoner confirmed
that Helen Kroger had been offered about twenty houses but he
had made the final decision, visiting the property with her. Lonsdale said he had intended to meet Houghton in January 1961 and
then seek permission from Moscow to go abroad to give the KGB
time to take steps 'to find out whether in fact Houghton was being

investigated'. He confirmed that his having a mistress in Brussels would be 'severely frowned upon' by Moscow. Lonsdale explained he was unwilling to reveal who he really was to prevent the British identifying his family and for fear they would publish his name, which 'would finish him in the eyes of the Soviet government'. The agent then imposed an additional condition as part of any deal with the British: that if he returned to Russia to be with his family and detected he had fallen 'under grave suspicion', he would be exfiltrated, given British nationality and money to live. Lonsdale talked at length about his state of mind, referring, said Elwell, 'somewhat bitterly to the intelligence racket in which he had become involved', from which he could not resign because he would not be able to find 'another job of comparable calibre'. Though, in the long term, capitalism was doomed, he enjoyed some aspects of it while being irritated by certain facets of life in the Soviet Union. He had become 'neither fish, fowl, nor good red herring, an alien both to the West and to the East'.[9]

Lonsdale's refusal to reveal his identity spurred Elwell's quest to discover the real person behind the 'legend'. The microdot letters to Lonsdale from his family had provided the first clues, but others slowly emerged. A letter Lonsdale sent from prison to Molly Baker (MI5, of course, intercepted and copied all correspondence sent by and to the five Portland spies while in prison) revealed the birthday of his young son, Trofim, born on 19 April 1958. At the end of the criminal trial a London newspaper published a photograph supposedly of the young Gordon Lonsdale with his mother. Further interviews conducted by the FBI in Washington State with relatives of the real Gordon Lonsdale confirmed that the woman in the photograph was definitely not his mother, and they later provided other pictures of the real Gordon Lonsdale as a baby with his mother as proof.

The source of the next clue was prison. In Wormwood Scrubs in May Lonsdale somewhat naïvely befriended a fellow prisoner who spoke Russian and offered to help pass messages out of prison. Lonsdale gave him a scrap of paper with his mother's address in Zubovsky Boulevard in Moscow. The prisoner went directly to the jail authorities with the information. MI6 made some enquiries, but no progress was possible because Lonsdale's genuine Russian name was still a mystery.[10]

The breakthrough came in June, as so often by revisiting stale enquiries with fresh eyes. Elwell recalled a bugged conversation between Lonsdale in his apartment and a girlfriend before his arrest in which he talked about being educated in California at an American high school with a good name. Elwell linked this detail to a follow-up interview he had conducted recently with another girlfriend of Lonsdale's, who recalled that Lonsdale had told her he had been brought up by an aunt in San Francisco and commented very knowledgeably about the city while watching a film. Back in December 1960, the FBI had investigated Lonsdale's supposed education in California but no school had a record of a former student named Lonsdale. Around June 1961 the FBI decided to plough the same educational ground, partly to establish if Morris Cohen had ever worked as a teacher in California, but also to ask schools if they recalled any Russian pupils from the mid-1930s who might be the elusive 'Gordon Lonsdale'. They struck lucky. The A to Zed School in Berkeley, California, had been founded in 1907 by the remarkable writer and educator Cora Lenore Williams, and pioneered instruction for children in small groups. The former principal of the school informed the FBI that: 'the only Russian who attended this school was one Konon Molody[11] who was born on 17 January 1922. He enrolled in the school in September 1936, and left after three weeks of the spring semester in 1938, stating that he had to return to Europe. He lived with an aunt, Tatiana Piankova, a dancing [ballet] teacher in Berkeley.'

Further research confirmed that Molody had previously attended the Frances Willard Junior High School in Berkeley, the first ever junior high school in California, for just over two years starting in March 1934. The FBI immediately started an 'intensive and expeditious investigation' to establish if indeed Konon Molody was Gordon Lonsdale, and they asked MI5 and 'Broadway' (MI6) to assist.[12]

Within days Hoover's men had assembled an outline of Molody's time in California and his family. Molody's mother in Russia had been a well-known doctor and had sent her son to the United States 'for educational purposes to escape communist influence'. He had lived with his aunt, Tatiana, and her daughter, Nina. Another aunt, Anastasia, also resided in California, in San Francisco. Molody obtained a US passport in 1938, but in 1939 he returned to the USSR, parting from Tatiana, Anastasia and a close friend, Irina Semech-enko. Tatiana and her daughter were reported to be living in France, Semechenko in Los Angeles, and soon the Bureau were tracking them down. The passport photograph of young Molody established beyond doubt by mid-July that he was identical with Lonsdale. The Bureau planned not to interview any of Molody's relatives until MI5 had confronted Lonsdale with the fact that Western intelligence now knew his true identity, in the hope that he would reveal everything he knew about his relations and their activities on behalf of the KGB. MI5 disagreed strongly. They believed Lonsdale's 'likeliest reaction would be to stay mum': the FBI and the Security Service should first assemble as complete a picture as possible of Lonsdale and then confront him. The stage was set for a tussle between the FBI and the Security Service.[13]

The Home Office continued to be unbending in its attitude to Lonsdale. 'It would be quite impossible to strike anything in the nature of a bargain with him,' the Permanent Under Secretary at the Home Office underlined to Hollis in mid-July. They did, however, agree to Lonsdale meeting the Krogers in the presence of MI5 in Birmingham prison: part of the plan of MI5's counter-espionage department to encourage them to cooperate. Elwell passed this news to Lonsdale when he met him at Strangeways prison at the end of July. He again brought a bag of cherries. Lonsdale said he was 'a little nervous in case the Krogers should find out how far he had already committed himself in conversation with' Elwell and how much he had already told him. He did not wish to return to the Soviet Union and be accused of betrayal. To encourage the Krogers to cooperate, Lonsdale said he would tell them that he had agreed to answer various questions about his work as an illegal in exchange for an early release. The MI5 officer again raised the issue of why Lonsdale was so reluctant to reveal his true identity, and the spy said that the Security Service might publish it. If so, he explained, 'it would immediately be assumed in Moscow that he had revealed everything'. And if he betrayed the name of another Russian spy, 'he would be recorded as a traitor and his name put down on a list of those who would be assassinated'.[14]

Careful plans were laid for the meeting between Lonsdale and the Krogers in Birmingham prison. Lonsdale was driven in a police car from Manchester and returned to prison there the same day,

15 August. Elwell carried a suitcase fitted with a hidden microphone so the meeting could be recorded. The transcript that resulted is an extraordinary document: the reader has the impression of being in the room with the participants. Lonsdale dominated the conversation, with Elwell and the Krogers interjecting only occasionally. The Russian spy repeatedly told the Krogers that they should consider giving MI5 information in exchange for a reduction of sentence or exchange with British agents in Russian hands. He stressed that it was their decision, but that they should not reveal the names of any active KGB agents, and if they sealed a bargain with the British they would still need to 'account for their actions' if they returned to Russia. It was agreed that the Krogers would ponder their response and Elwell would visit them in a week or two to find out their decision. Afterwards the MI5 officer met Lonsdale alone. Now that he had discussed with the Krogers the possibility of passing information to the British, Lonsdale said he had 'burnt his boats' and there was 'no going back'. What 'he had done, he had done with open eyes'.[15]

Worldwide, potential leads about Lonsdale's espionage were followed up laboriously by Western intelligence agencies. The primary focus over the summer of 1961 was to complete the jigsaw of Konon Molody's life. The dispute between MI5 and the FBI about when to confront Lonsdale in prison with his true identity was resolved by the Bureau retreating, and starting to trace and interview Molody's relatives. These included not only Molody's two aunts who had lived in California, Tatiana and Anastasia, but a third aunt, Serafima, and her son − Molody's cousin − then living in Berkeley. By 1961 the three sisters had settled thousands of miles apart: Tatiana in France, Anastasia in California and Serafima in the USSR. One abiding mystery was the identity of Molody's mother. According to stories told by Tatiana to some relatives, Tatiana was not Konon Molody's aunt but his mother. She had made no mention of a son, however, when she had applied for a US passport in 1932.

In mid-July, a Soviet expert in D1 and colleague of Elwell made an exciting discovery. Now armed with Lonsdale's real surname, Molody, he searched in the Moscow telephone directory against the street name given by Lonsdale to the fellow prisoner in Wormwood Scrubs, Zubovsky Boulevard, where the spy said his mother lived. He found that a lady called Yekaterina (or Evdokia) Molodaya was living at 16/20 Zubovsky Boulevard, who was clearly Molody's true mother. Evdokia was a missing fourth sister. Elwell sketched out a family tree to help research. Especially intriguing was testimony from Molody's cousin in Berkeley that his mother and his aunt Tatiana became nervous and upset whenever he asked about Konon Molody. They were clearly concealing something.[16]

Part of the mystery was solved by Aunt Tatiana when interviewed by the FBI in Paris in the autumn of 1961. Tatiana ran a ballet company based in the French capital and travelled extensively. A meeting finally took place in the American embassy on 14 September. It began at 7 p.m. and lasted almost five hours. Tatiana Piankova was born into a wealthy Russian family whose money had been confiscated after the 1917 Bolshevik Revolution. There had been seven children, and in the chaos following the revolution they had dispersed. Tatiana and Anastasia escaped from Russia with one brother, travelled through Japan and China and finally emigrated to the USA, where they settled in California. Serafima had left to live in Estonia, while two other sisters, Manya and Evdokia, stayed in Russia with a second brother.[17]

In 1914 Evdokia married Trofim Kononovich Molody, a prominent scientific writer and editor in Moscow, who originally came from eastern Siberia. Evdokia had trained as a doctor and practised in Moscow, where she gave birth to two children, Natalia in 1917 and Konon in 1922. Trofim died of a brain haemorrhage in 1929. The mother was obliged to bring up the two children alone, and in desperate circumstances. That year Stalin proclaimed the end of Lenin's New Economic Policy, and the beginning of forced industrialisation in the towns and collectivisation in the countryside. The result was a catastrophic drop in food production. Within a year or two, famine stalked great tracts of the Soviet Union and millions perished. Meanwhile, in California, Tatiana and Anastasia were living tolerably despite the Great Depression. In 1932 Tatiana decided to go to Moscow to visit her elderly mother and her sister, Evdokia. 'At this point in the interview,' the FBI note recorded, 'Mrs Piankova became quite emotional and agitated and began to cry.' In Moscow young Konon 'appeared to be in extremely poor health and suffering from malnutrition'. When sent out to collect rations for the family, he often ate them all before he returned home. Tatiana asked if she could bring Konon to the USA for a better life.

Evdokia agreed. Tatiana returned alone to the USA, planning to obtain an entry visa for Konon by representing him as her adopted son. This proved impossible. Tatiana and Anastasia agreed that the only sure way was for Konon to appear to be Tatiana's own son. Tatiana therefore declared Konon to be her child by a former marriage, whom she had been forced to abandon in Russia when she fled to America after the 1917 Revolution. The visa was granted and Konon travelled to California to live with her and her sister, Anastasia. At school he was an outstanding student and he had Russian lessons in his spare time. Aunt Tatiana also told how Anastasia travelled back to Europe with Konon in 1938, leaving him to study in Estonia, but he travelled on to the USSR the following year to join the Russian army. She had heard nothing from him since. Tatiana revealed that she went back to Russia in 1959 to visit Evdokia, but she 'absolutely refused to discuss Molody at all'.

Elwell and one of the FBI's men in London, John Minnich, pored over the Tatiana Piankova interview and agreed there were numerous anomalies and inconsistencies. She was clearly not telling the whole truth. They travelled over to Paris together on 2 October. Minnich interviewed Piankova again that night. Haltingly, she revealed more missing pieces of the jigsaw of Konon Molody's life. He had been married twice, the first time to a woman whose father was a top official in the Tupolev aircraft factory in Russia, and with whom he had his daughter, Lisa. Back in Moscow in 1933, it transpired that Molody had only been allowed to leave the Soviet Union because of the intervention of the wife of Maxim Gorky, the renowned Russian novelist, at the relevant Soviet ministry, and with the help of her beautiful daughter-in-law, who was the mistress of Genrikh Yagoda, the head of the KGB.

Assistance came also from a sympathetic Russian Orthodox priest. To enter America, young Molody's papers required a false birth certificate stating that he was Tatiana's natural son. Desperate to help her beloved son, Evdokia begged the priest for assistance and he furnished the fake certificate. The catalyst for Molody returning to Russia in 1938 had been a letter from his mother in Moscow, telling him he must decide then 'whether he wanted to be an American citizen for the rest of his life or to return to Russia'. After a month's reflection Molody said he wished to go back to Russia. His aunt Anastasia (whom Molody detested) was 'infuriated' and organised a trip through capitalist countries in Europe 'so he would always remember his mistake in returning to the poverty of Russia'. The ruse failed. At the end of the tour Molody, aged only sixteen, took a train from Estonia to Moscow.

Tatiana disclosed to the FBI that she had recently written to her sister, Evdokia, in Moscow to tell her that the FBI had suddenly been taking a very great interest in Konon Molody. If they were not aware before, the KGB now knew that Western intelligence services had discovered the true identity of Gordon Lonsdale. Elwell met Tatiana alone the following evening in a café on the Boulevard Saint-Germain. She was, Elwell wrote, 'a thin, little woman whose plain, worn features are redeemed by bright intelligent eyes. She was dressed in black and had over her shoulders a fur cape.' He confirmed that he had met her nephew, Konon Molody, on several occasions, both before and after he knew his real identity. She said, 'with fervour she would do anything for [Molody] because she was very fond of him and indeed in some ways preferred him to her own daughter'. She asked Elwell when he saw Molody next to tell him that 'she did not consider that he had done anything wrong or dishonourable'. All the Molody family secrets which Tatiana and her sisters had been concealing for almost forty years had finally been unlocked.[18]

The follow-up meeting with the Krogers in Birmingham prison after their reunion with Lonsdale on 15 August was, in Elwell's words, 'rather unsuccessful'. They were not able to give a decision, saying they expected Lonsdale to be present, and raised the question of what guarantee the British could offer that they would honour any bargain. Elwell met Lonsdale a few weeks later on 19 October in Strangeways prison. The Russian illegal 'readily undertook to do anything he could to get the Krogers to play their part in any agreement that was reached'.[19]

A few days before that meeting Elwell had drafted a considered but calculating note, setting out his thoughts on how best to persuade Lonsdale to cooperate. His main point was that 'the sooner Lonsdale makes up his mind not to return to Russia, the better for us'. Elwell reasoned that the best way to achieve this aim was for the Security Service, in the greatest secrecy, to engineer a leak to the press of Lonsdale's true identity, preferably in America. Once his name had been revealed, it 'will have been used in anti-Soviet propaganda and [Lonsdale] will be held responsible [by the KGB]'. The British should only make an offer to Lonsdale of a reduced sentence after the leak had occurred and the spy understood that not only MI5 knew his true identity, but that the KGB knew the Security Service knew. This, Elwell argued, would place the maximum pressure on Lonsdale. Elwell's suggested terms for a bargain were a reduction of Lonsdale's prison term from twenty-five to fifteen years, and the Krogers' from twenty to fifteen.[20]

The plan to leak Lonsdale's identity was put to the FBI by the Security Service, proposing that the leak would seem 'most natural' if it took place in the San Francisco area. A terse and factual statement to be given to selected media outlets in California was agreed by the FBI and MI5. The final version read:

> Attorney General Robert F. Kennedy today announced that Gordon Arnold Lonsdale, arrested in England last January by British authorities for espionage and now serving a twenty-five-year prison sentence there, is in reality a Moscow-born, thirty-nine-year-old Russian national named Konon Molody. FBI investigation determined that Molody came to the United States in 1933 posing as his aunt's eleven-year-old son. He resided with her in Berkeley, California, until around 1938 when he returned to Russia.[21]

The leak took place on 24 November and was discreetly reported in various newspapers. The *Daily Telegraph* in London described the announcement that Lonsdale was in fact a Moscow-born Russian as the 'first official confirmation of what has been assumed since Lonsdale and his four accomplices were exposed'.[22]

Elwell began the next meeting with Lonsdale on 28 November in Strangeways prison by saying, 'I'm afraid you must be rather upset.' In response the spy looked alarmed and puzzled and said, 'Why?' Elwell showed him a selection of newspaper cuttings confirming his true identity. Although 'distinctly ill at ease', Lonsdale's reaction was 'sombre indifference' to the news, arguing that the revelation of his real name 'could not possibly make any difference', either to his position or that of the British government. He then badgered Elwell about when the British would make their offer to him. The MI5 officer pretended 'to be despondent about the possibilities and said the question was still being discussed'.[23]

In reality, the offer to Lonsdale was being finalised. The Home Office's supposedly trenchant opposition to a 'bargain' had dissipated over the summer and been replaced by the realpolitik of espionage. A detailed list had been developed of the operational intelligence to be obtained from Lonsdale and the Krogers if they agreed to cooperate. The offer made its way up the hierarchy of the British establishment in hermetic secrecy. Having been approved by Hollis, it needed not only judicial but political sign-off. There was considerable resistance and discussion. The deal was finally agreed by the Lord Chief Justice and the Home Secretary, and then by Harold Macmillan himself towards the end of November. Naturally, there was much concern about how public opinion and the House of Commons would react if or when the deal became public and about the precedent it might set. The offer was based on Elwell's proposals: Lonsdale's sentence would be reduced from twenty-five to fifteen years, and the Krogers' from twenty to fifteen years, in return for answering a questionnaire about their espionage and Soviet intelligence.[24] Elwell produced the official document from his briefcase on the morning of 6 December in Strangeways prison and presented it to Lonsdale. The spy read it slowly and deliberately and then returned the paper to him. He declared, 'Well, I'm not interested,' adding that it was 'an insult not only to my intelligence but to my common sense'. Lonsdale explained that it was not worth risking his life for a reduction of only ten years; he gave Elwell the impression that he was thinking in terms of release after a maximum of three years, so that 'he still had friends in responsible positions in the KGB' on his return to the Soviet Union. Elwell stressed that the offer was final and that there was no question of negotiation. Nonetheless, although no doubt deeply disappointed at Lonsdale's response to the offer after his immense effort in fathering it, Elwell stayed with the prisoner for several more hours. Lonsdale remarked that he was 'a little concerned' that his true identity had been revealed in case it 'got his family into trouble', and talked about his past life (describing an interview with Lavrenti Beria, Stalin's Secret Police chief, when Lonsdale 'literally shivered with fright').[25]

In the weeks that followed, Elwell put forward an equivalent offer to the Krogers. The MI5 note of that meeting is still classified, but the two KGB agents said they wanted another meeting with Lonsdale to discuss it. In January 1962 Lonsdale agreed to advise the Krogers if asked. He discussed the intelligence questionnaire with Elwell, commenting that 'the Russians did not use microdots very much as they regarded them as insecure and inefficient', and Lonsdale teased Elwell about his coded messages, presuming that the British 'had been able to decipher all his messages. If we [MI5] had not succeeded, then Heaven help us.' Lonsdale went on to suggest he had been 'very much more active than most' illegals, the majority of whom were inactive in peacetime, and had always thought that 'there was about a 25 per cent chance that he would be caught'.[26]

The first nine months of 1962 slid by with only one significant development in the Lonsdale case.[27] In the spring the FBI passed to MI5 new information about Lonsdale gleaned from a KGB officer called Anatoli Golitsyn. In December 1961 Golitsyn had defected to the CIA in Helsinki and it had taken several months to debrief him. When interrogated about Russian illegals, Golitsyn spoke of a young boy who had spent part of his childhood in a port city like San Francisco in America and later returned to the Soviet Union, where he was recruited in about 1950 by the KGB and placed in the Illegals Directorate. While in the KGB, Golitsyn had seen a group photograph of the young boy at school in America, and had later observed him in Moscow in 1951. When shown a 1961 photograph of Lonsdale, Golitsyn identified him as Konon Molody.[28]

The second meeting between Lonsdale and the Krogers took nine months for the Security Service to engineer. The Krogers were driven down from Manchester to Birmingham on 27 September 1962. After handing out fruit, chocolate biscuits and cigarettes, Elwell again produced the British offer to the Krogers. Lonsdale said: 'it was entirely up to them whether they accepted it or not'. He pointed out that their position differed from his: they were older, and not intelligence officers. The response from the Krogers was swift and uncompromising. 'Kroger said with emphasis,' wrote Elwell, 'and with the truculent tones that he always adopts when he is talking of things related to his work in the RIS, that he was not the slightest bit interested in our "offer".' His wife agreed. Elwell noted that the Krogers were 'bitterly anti-American and hopelessly blinkered communists for whom nothing that is American is good'. Elwell doubted whether they would cooperate unless the British offer were 'made very much more attractive'. Elwell concluded that 'So far as

Lonsdale's morale is concerned . . . I should say that it is as high as it has ever been.'

The British government was unwilling to be more generous, but the existing offer was insufficiently tempting to persuade Lonsdale or the Krogers to accept. A deal was probably impossible. There is no published case of a convicted spy who has agreed to provide sensitive intelligence about their service in exchange for a reduced prison sentence, been released early and then either returned to the country of their intelligence agency or remained in the country where they had been imprisoned. The result with Lonsdale and the Krogers was stalemate – a stalemate to be broken only by events outside the UK, and outside the control of the Security Service.

II

The balance of power in the Lonsdale case between London and Moscow tilted suddenly in favour of Russia in November 1962, when a British businessman, Greville Wynne, was arrested in Hungary. He was flown to Moscow in great secrecy and imprisoned. Wynne had in fact been working for MI6 for some time and been the courier for the Russian spy Oleg Penkovsky, who had been passing immensely important intelligence about Soviet nuclear weapons to the West during the Cuban Missile Crisis. Penkovsky had been betrayed shortly before Wynne's arrest and it was he, under harsh interrogation, who had divulged Wynne's name to the KGB. News of Wynne's fate soon reached London.

The first straws blown by the Soviet wind were two typewritten letters in English sent within a couple of weeks of Wynne's arrest to the barrister who had defended Lonsdale at the criminal trial, William 'Barry' Hudson. The first was from a lawyer based in Warsaw, Mieczysław Buczkowski, who stated that he had recently been approached by two women living in Poland, Halina Lonsdale and Eugenia Panfilowska, who he said were the wife and mother-in-law of Gordon Lonsdale. They had requested that he assist them in exchanging letters with Lonsdale and 'help them in other matters concerning this case'. Enclosed was a bland but effusive letter from Halina to Lonsdale passing on family news and asking him to write. This correspondence was duly passed on to Lonsdale in prison, with

MI5's approval. Although the Polish lawyer claimed that he had found Hudson's name in press cuttings about the case, a more likely explanation is that Hudson had discussed the case at a party with a diplomat at the Soviet embassy in London several months before – a conversation which had attracted the interest of the Security Service and which led to Hudson promising to pass on to them details of future contacts with Soviet diplomats.[29]

MI5 understood only too well that the KGB were starting man-oeuvres for a possible spy swap. On the evening of 11 January 1963, the Home Secretary and Hollis briefed the Prime Minister personally on the case. Macmillan clearly shared the frustration of the Security Service over Lonsdale's obdurate refusal to let down his guard, but was still intrigued by the possibilities of some form of deal. He wrote in his diary:

We have in our prison a Russian spy, Lonsdale, sentenced to 20 years [in fact 25] imprisonment by the L.C.J. He suborned a num-ber of minor characters (British subjects) who (as such) are guilty of treachery to their country. L. is said to be a good 'professional' intelligent officer of 20 years standing, agreeable, clever, and of good character. *If* he wd. talk, he cd. give us information of real value. But he *won't* talk. We have offered substantial reductions (of five years or so) but he is not tempted. If he has heard of the capture by the Russians of one of our men – Wynne – he may be hoping for an exchange. Of course, in value there is no com-parison between the pieces – one is a pawn (ours) the other a knight or bishop (the Russian). But public opinion *might* force us to a disadvantageous swap. If L. talks, it wd be worth reducing his sentence to the end of (say) 1963 or 1964. But will he talk? The best thing is that he talks (in exchange for early release) and *afterwards* in exchange for Wynne. This would enable him to go back to Russia. If he talks and is released, he cannot. It is quite an interesting position, described to me last night by the authorities chiefly concerned – Home Office and MI5.[30]

In early 1963 Elwell was moved to other counter-espionage work. Handling of the Lonsdale files was passed to Freda Small, another example, like Evelyn McBarnet, of how MI5 was finally beginning – albeit in a minor way – to recognise the abilities of women in the

Service. Letters continued to arrive from Poland for Lonsdale and he was allowed to write some letters to his family in response.[31] Furnival Jones told Hollis at the end of February that it seemed 'clear that exchange is in the wind'.[32] MI5 watched and waited. Lonsdale did the same. In May 1963, Wynne was convicted in Moscow of spying and sentenced to eight years in prison. Penkovsky was executed. Within days of the trial a clutch of articles appeared in British newspapers questioning whether Lonsdale would be swapped for Wynne.

After Macmillan's resignation as Prime Minister in October 1963 due to ill health, top-secret discussions about a spy exchange gathered pace at the highest levels of the British government, partly influenced by a CIA debriefing of a returned American prisoner who revealed that, already by summer 1963, Wynne had lost more than two stone in weight. In December the Lord Chief Justice indicated to Roger Hollis and the senior official at the Home Office there would be no judicial objection to an exchange. Henry Brooke, the new Home Secretary, was sceptical. The Prime Minister, Alec Douglas-Home, debated a possible exchange with senior ministers in January 1964 and it was agreed that such a swap was not 'appropriate . . . at present' because of the 'difficulty that would be experienced in making a convincing public presentation on the exchange'. A visit to Moscow by Wynne's wife at the beginning of March 1964 changed the British government's calculus. She wrote a private letter to the Foreign Office expressing alarm at his bad treatment in the Soviet prison and the serious deterioration in his health. Having seen the letter over the weekend of 14–15 March, the Prime Minister was persuaded that 'on humanitarian grounds, there was now no alternative but to work towards' an exchange. There was only one recent precedent for a spy handover and that involved the Americans two years before, when Gary Powers was swapped for Willie Fisher. The British had no experience. Top government officials drafted lengthy memoranda discussing whether Wynne and Lonsdale should take off simultaneously in separate planes from Moscow and London, or be exchanged at an agreed location abroad. Roger Hollis and Sir Dick White, the head of SIS, were consulted.[33]

The focus switched to negotiating in the strictest secrecy with the Soviet Union how and when the exchange could take place. On

17 March the British ambassador in Moscow was told of the decision by top-secret cable and instructed to speak immediately to the Russian Foreign Ministry. The first reaction of the acting head of the Second European Department, Comrade Miranova, astounded the ambassador: it was 'to ask who Lonsdale was'. When this was explained, Ms Miranova said she would 'report to the leadership' and get back in touch when there was a decision. The Soviet reply came three weeks later. The exchange could be effected 'at any time convenient to the British side', replied the Russians, through a simultaneous handover at a crossing point in Berlin, Warsaw or a Polish port. To eliminate any risk of the Russians reneging on the bargain, the top British officials recommended a simultaneous exchange in Berlin. Molody would be flown out clandestinely by military plane to avoid media attention and Wynne returned to the UK by the same means to give the British 'a chance to brief Wynne before he meets the press' and underline that he should not confirm that he was connected with MI6 (or 'our friends', as the Cabinet Office correspondence coyly describe them). Lord Home agreed, but revealed an abiding distrust of the Soviets: 'Might they give him [Wynne] some slow-working poison or try and kill him here, or do they not mind the stories of prison treatment which he will tell? Anyhow go ahead. I think we ought to be quite ruthless with the press and tell them that if they leak there is every chance he will be killed.'

In Berlin the Heerstrasse checkpoint was agreed on for the exchange on 22 April, and the British and Russians nominated special phone numbers in Berlin so both sides could liaise in the forty-eight hours before the swap. Ministers were apprehensive that the newspapers would sniff out the story and jeopardise the exchange, so elaborate arrangements were made to keep Molody's release as covert as possible.[34]

At 10.15 a.m. precisely on 21 April 1964, two unobtrusive cars hired by the Security Service drew into Winson Green prison in Birmingham. Five men clambered out: a Home Office official, the Deputy Chief Constable of Birmingham, an MI5 officer from Operations, George Clayton, and two police officers. Molody had only been told five minutes before, when he had been summoned from the bathhouse, that he was to be moved, and was instructed to collect his belongings and change into his own clothes. When handed his conditional release by the Home Office official in the prison governor's office he resisted at first, raised his voice and did not want to leave, a 'deeply shaken, a trembling, deeply shocked man' according to Clayton.[35] Perhaps it was the suddenness of the move, perhaps alarm at the prospect of execution back in Moscow. Moments later, however, his resistance evaporated and he became cooperative. Lonsdale was handcuffed to one of the police officers and sat in the back of the lead car with the deputy chief constable on his other side and Clayton in the front seat. At 10.30 a.m. – only fifteen minutes after the two cars arrived – they roared off to the M1 motorway towards London. Within minutes Lonsdale had regained his usual good humour and the deputy chief constable ordered the handcuffs to be removed. The police officer had 'for a moment a little difficulty in unlocking them and [the deputy chief constable] said that probably Lonsdale would be better at it. Lonsdale replied that he was afraid that he was rather out of practice; it had been a long time since he had learnt about locks.' This sally set the tone for 'an animated conversation . . . in the car – Lonsdale setting the pace' for the remainder of the journey to the Royal Air Force base at Northolt in north-west London. Lonsdale sat between the two taller men on either side, 'his almost black hair, rather boyishly untidy, falling in a lock over his right eye around a rather sallow but full, merry face . . . [H]e looked healthy enough, though he was dressed in a poor-looking light-blue suit and his shirt, although clean, did not quite meet round his neck beneath his tie . . . He frequently emphasised what he was saying by beating on the back of my seat with his hands.'

Lonsdale chatted away, giving his views on: prison (one of the worst features of his stay in Birmingham 'was the lack of intelligent people to talk with' and the 'filthy' lavatories); how poor his defence was at the trial (he wished he had defended himself); his job as a spy (no more exciting than that of a 'trials driver or a test-pilot, but people and the press always made these things out to be glamorous'); and the KGB ('the Russians could be suspecting him for no reason at all and he would not know it. To be under suspicion in Russia was a worse crime than the crime itself'). The MI5 officer detected a note of uncertainty underlying Lonsdale's bubbling energy: 'He spoke quite a lot about his family but curiously unlike a man about to be reunited with them after so long – or did he have his misgivings? . . . The longest he had ever been with his wife was three months . . . On another occasion he said, "I wonder where it will all end."'

They met an RAF liaison officer where the M1 entered the outskirts of London and the cars threaded their way through the suburban landscape of Watford, Rickmansworth and finally Ruislip. 'Lonsdale commented that we were in country which he knew very well – were we taking him to 45 Cranley Drive? In fact, by sheer fluke, we passed the end of the road, noted by Lonsdale and pointed out by him with great amusement . . . Lonsdale said he chose this district himself because of the large number of American transmitters in the vicinity.' There were only a handful of people around the aged RAF Valetta transport aircraft as the cars pulled up beside it on the tarmac at Northolt a couple of minutes after 2 p.m. Ushering Lonsdale into the aeroplane, the MI5 man noted 'a sudden and marked return of nervousness' as he turned and left him in the hands of a new escort from MI6. 'His few bits and pieces were put on board, the door closed and, within five minutes of our passing through the gate, the engines had started and the Valetta taxied towards the runway, waiting for its scheduled take-off time. It took off at 14.15.'

After a three-hour flight Lonsdale landed at the RAF airfield at Gatow in the British sector of West Berlin. Although it was inevitable that rumours of a spy swap started to buzz within hours of Molody leaving prison in Birmingham in civilian clothes, the British media only winkled out the story late that night. The senior Foreign Office News Department official was at the Japanese embassy in London at 10 p.m. when he was phoned by the correspondent of the *Daily Express* who specialised in spy stories, Chapman Pincher, 'obviously in a restaurant or club . . . [who] said he was having to use a coin box machine'. Pincher declared that he was going to file a story that 'Lonsdale had been released from his British prison in the morning and taken abroad (probably to Berlin), that this meant that an exchange for Wynne would probably be taking place in Berlin at any moment'. The Foreign Office newsman asked Pincher to 'hold [the story] up at least until his second edition so that it did not become the big story in all the press', and 'guaranteed that there would be a government statement as soon as this was feasible'. News of the exchange was splashed across the front page of the second and later editions of the *Daily Express*. By this time the story had leaked to the international news agency, U.P., and within an hour or two the Foreign Office was 'inundated with enquiries'. There was such consternation in Whitehall about the *Daily Express* story that MI5's legal adviser was ordered to carry out an urgent inquiry as to how the newspaper discovered details of the Molody transfer arrangements. He concluded that nothing had been leaked inappropriately: the information had been unearthed through diligent journalistic sleuthing by the Birmingham office of the *Express*.[36]

On 22 April Molody was woken around 4 a.m. in the military prison at Gatow. As a grey dawn suffused through the mist, he was driven through the deserted Berlin streets in a black Mercedes Benz until they reached the road for Hamburg. Even early in the morning the highway was normally humming with convoys of long-distance lorries, but that day the tarmac was empty: it had been closed to all traffic until the exchange was over. After passing through the Heerstrasse checkpoint, the car braked to a halt in a lay-by in No Man's Land at 5.25 a.m. Ahead lay a forlorn stretch of road, and at the centre point of the Heerstrasse checkpoint a short white line had been painted. Beyond lay the German Democratic Republic, under the influence of the Soviet Union. There a barrier lifted and two cars drove forward. From the second emerged Wynne. A KGB officer who had worked with Molody, and one from MI6 who knew Wynne, confirmed the identity of their men. For a moment the two groups faced each other across the white line. The Soviet consul rasped out, 'Exchange' in English and Russian. Molody and Wynne passed each other on the tarmac and were bundled into cars, which sped away with their human cargo.[37]

9

The Krogers

On 30 April 1964, exactly eight days after his exchange, Lonsdale wrote his first typewritten letter to Helen Kroger in Styal prison in Cheshire. He addressed her as 'Sarge' and signed the missive with one of the names by which he was known to the Krogers, 'Arnie'. The letter was sent from an address in Warsaw, referring to his fictional Polish wife, Halina, and mother-in-law. Lonsdale said he was 'still in a daze' after 'the overwhelming . . . speed' of his release, urging Helen to 'be patient while being sure that everything possible is being done'. At the same time he started to correspond with Peter Kroger, then back in Wormwood Scrubs for surgery on his right hand (he suffered from a complaint which made the tendons of his palms contract, drawing the fingers into a fist). Soon letters for Helen Kroger began to arrive also from a long-lost 'Polish cousin', Maria Petka. All these letters were the prologue to an extended correspondence between the freed KGB spy and his fellow agents. It was orchestrated by Moscow – as the Security Service understood only too well – to bolster the morale of the Krogers and to lay a trail of evidence suggesting that they had Polish citizenship, so helping to stymie any potential extradition to the USA.[38]

His visits to the Krogers in the early days of their imprisonment had confirmed to Elwell that, unlike with Molody, he had no hope of turning them. Although his last meeting with them and Lonsdale had ended in chilly politeness, Elwell later recalled Kroger as a 'horrid man' and his wife as 'even more horrid'. They were committed Soviet communists and had taken weighty risks to work undercover as illegals, like Molody. They had however, unlike Molody, betrayed their country rather than make dangerous sacrifices for it. Although he despised Soviet communism and the shackles it imposed on human freedom, Elwell despised anyone who betrayed their country more – and even more so if, as with Houghton and Gee, it was for money.[39] As communist idealists, the Krogers in turn must have scorned Elwell in his pinstripe suit and bowler hat as a privileged representative of corrupt capitalism, and made relatively little effort to conceal their contempt. Sometimes, too, they could be pompously self-righteous. One day in prison Helen Kroger had a conversation with the governor. He said he understood why working-class people might become communists, but not people from the British upper class like Philby, Burgess and Maclean. Po-faced, she retorted that he had clearly not read Karl Marx: 'Marx said that the best sons and daughters of any country, if they love their people, will do all they can to bring about socialism.' The Krogers, unsurprisingly, refused to be interviewed by either the FBI or the CIA.[40]

Some friends and clients of the Krogers felt so betrayed by their duplicity that they refused to allow their names to be uttered in their presence.[41] Others, however, remained stoically loyal. Throughout their time in prison, for example, the Krogers were visited regularly by an antiquarian bookseller friend, Freddie Snelling (who also corresponded with Lonsdale). After George Blake's daring escape from Wormwood Scrubs in October 1966 (which was not organised by the KGB but by his cellmate and two anti-nuclear peace activists – although this was not known at the time), prison security for the Krogers was tightened sharply. Both were transferred to maximum-security institutions: Peter to Parkhurst prison on the Isle of Wight and Helen back to Holloway. For a period they were placed in solitary confinement and not allowed to touch letters sent to them in prison: a guard held them up and they had to read them at a distance. Peter Kroger was a popular prisoner, however, helping to teach other prisoners to read and write, and writing letters for them. Through Snelling he ordered a constant stream of serious books which helped strengthen his personal commitment to communism. According to Kroger's friend, his preference was for 'weighty, two-volume histories, while any new work whose title bore words like ... "dialectical materialism", or "metaphysical contradictions" drew him like a magnet'. A self-disciplined man, he gave up smoking. Helen Kroger, on the other hand, was unable to do so, and once she had smoked through her ration would chew a piece of orange peel. Snelling supplied her with cigarettes and a weekly bunch of flowers.[42]

During the months while the Cohens were first in prison, Elwell had continued with his investigation into how they had obtained their false passports in the name of Kroger. He liaised with the New Zealand Security Service (NZSS), which started a major investigation. MI5 in London was sent copies of all the documents submitted by Morris and Lona Cohen to the New Zealand legation in Paris to obtain their false passports in 1954 and then renew them in 1959.[43] In May 1961 Hollis reassured the Romer Inquiry that in his opinion the New Zealand authorities had behaved quite properly in issuing the passports to the Krogers.[44] Behind the scenes, however, evidence had already emerged which was throwing doubt on this assessment. The month before, a New Zealander working for a government minister in London had read in the newspapers at the time of the Portland trial about the Krogers obtaining their passports fraudulently. He remembered that he had worked in the New Zealand legation in Paris with two people who worked in the Passport Office there who he thought were communists. One was called Desmond Patrick (known as 'Paddy') Costello.[45]

The files in the MI5 Registry showed that Costello had in fact been on MI5's radar for some time, occasionally bleeping on the screen and then disappearing. A highly intelligent New Zealander and gifted linguist, he had studied at Cambridge in the 1930s. He became a lecturer at what later became Exeter University just before the war. By then he had married a woman called Bella Lerner, from a family of Russian Jews based in East London, who was a well-known communist. Costello was sacked because of links with a student at the university who was convicted in 1940 under the Official Secrets Act and because of his communist sympathies. He joined the New Zealand army and fought with some distinction before being accepted by the country's Diplomatic Service in 1944 and being sent to the newly opened New Zealand consulate in Moscow. The files showed MI5 had strong concerns about Costello's possible communist sympathies and reservations about him working as a diplomat. The British government exerted pressure on New Zealand to encourage Costello to find a job outside government service. Senior New Zealand diplomats proved reluctant to act against Costello when confronted with evidence of his possible treachery. However, while Costello was working in the consulate in Paris from 1950 as a first secretary and issuing passports, the British insisted he be discreetly excluded from access to sensitive material. In 1955 Costello finally found a job as Professor of Slavonic Studies at Manchester University. Despite some intermittent interest from the Security Service, he then disappeared off the MI5 radar screen until Elwell began investigating him in April 1961.

From his research in the Registry files, Elwell pieced together details of the brothers and sisters of Costello's wife, which suggested some tantalising but highly speculative parallels or even links with the Cohens. Both families, for example, had Russian-Jewish origins and were fervently communist, meaning that some of the Lerners had, like Morris Cohen, fought in the Spanish Civil War, or were involved with communist academics at the London School of Economics. One even had a connection with the antiquarian book trade. Nothing solid could be proved. At around the same time, MI5's Soviet counter-espionage section 'discovered that KGB agents were obtaining from Somerset House [the London archive of UK birth, death and marriage certificates] birth and death certificates of dead children, whose identities were under consideration by the RIS for illegal cover purposes'. Using these certificates of dead children, the Russians could obtain documents in the children's names to create false identities for their agents, who were known as 'dead doubles'. The Security Service put in place a system for checking applications for birth and death certificates to spot ones which might have been made by the KGB. Two such applications had been made by someone called 'B. Green' from an address in Exeter on the same day in December 1960. They appeared a little suspicious and were investigated, but there was no MI5 file on anyone of that name so the matter was dropped.

As the winter of 1961 approached, Elwell's investigation of Costello slowed to a halt: he and MI5 had limited resources and other priorities. MI6 in Paris made some enquiries in the autumn and found it 'very odd' that the Krogers bothered to travel to the New Zealand consulate in Paris in 1959 to renew their passports ('could it be that they had an accomplice of some sort in the New Zealand embassy there?' SIS asked rhetorically), and that Lonsdale arrived in France on exactly the same day.

For lack of further evidence, the investigation stalled until early 1963. By then, Elwell had moved to other work at MI5. In February there was a sudden flurry of renewed interest in the Costellos because of 'information from a Top Secret and reliable source that up to about two years ago Mrs Costello was an important agent of the Russian Intelligence Service'. The Security Service 'had no reason to suppose that she is not still an agent', and it was 'probable . . . that Professor Costello was involved with her'. This source, code-named 'Kago', was the Soviet defector Anatoli Golitsyn – the same source who had provided useful intelligence on Konon Molody in the spring of 1962. The Costello files were summoned from the Registry and an alert officer spotted a bizarre anomaly. The address in Exeter from which both the suspicious December 1960 birth certificate applications had originated was the one where the Costellos had lived until 1944, and there was a 'strong simi-larity' between this applicant's handwriting and that of Bella Costello. MI5's handwriting expert concluded in a top-secret memo that both documents were written by the same person, and a senior officer wrote: 'We therefore now have what practically amounts to written proof that in December 1960 Mrs Bella Costello was a KGB agent employed in an illegal support role. Experience suggests that she was probably a trusted agent of some years' standing to have qualified for such employment.' The Security Service guessed the Costellos were recruited in Moscow between 1944 and 1950. Since a number of former associates of Costello were still in government service, the implications for the New Zealand Security Service were potentially very serious, and its head was immediately informed. One of the most prominent of these associates was Balfour Douglas 'Doug' Zohrab, who was then Consul General and Permanent Rep-resentative of New Zealand to the United Nations Office at Geneva, but according to MI5 had previously been 'of security interest'. The Zohrab case was so important that the agreement of the New Zealand Prime Minister, Sir Keith Holyoake, needed to be obtained before investigating Zohrab's activities in Geneva and informing Swiss intelligence.[46]

The Costellos were placed under surveillance. Their phone was tapped and letters checked. In June 1963, at Ealing Tube station in west London, 'Paddy' Costello was observed to meet a known KGB officer called Vladimir Yermakov, who was based at the Soviet embassy, and they lunched together at a pub in Perivale. They both travelled on the same train back to London but took great care to sit in separate carriages: a frequent ruse to evade being tailed. MI5 drew the obvious conclusion: 'we have . . . identified Costello as a

Harry Houghton and Ethel Gee around 1960. Suspicion was aroused by him 'living above his means', like buying new cars, when earning only about £1600 a month (based on 2020 values).

Portland naval base, 1959. Houghton and Gee worked in the prominent UDE office building in the foreground on the left.

MI5 surveillance photo of Houghton and Gee in the lobby of the Cumberland Hotel, London.

The mysterious Gordon Lonsdale: London business-man and seller of jukeboxes with a Canadian passport, but no personal history before 1955.

Proof Lonsdale was a Russian spy: the hidden cavity in his cigarette lighter and the KGB cipher pads within, found by MI5 on 12 September 1960.

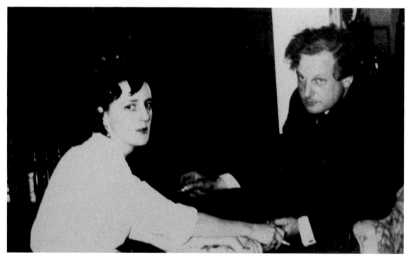

Spycatcher photographed by his target: Elwell in his London flat with family friend Elizabeth Oliver, unwittingly snapped by Lonsdale in 1956. When found in Lonsdale's secret cache in September 1960, Elwell himself briefly came under suspicion by MI5.

David Whyte on his wedding day in London, 10 September 1960. His marriage caused him to miss MI5's momentous search of Lonsdale's deposit box two days later.

A caricature of David Whyte in his typical office attire of bow tie and three-piece suit, drawn by one of his Security Service secretaries.

Charles Elwell in Royal Navy uniform before being captured in March 1942 by German forces in occupied Holland.

A drawing of Charles Elwell by a fellow prisoner in Colditz Castle during the Second World War. Elwell had been moved to Colditz after an escape attempt.

Ready for the Palace: Charles and Ann Elwell at home before the Queen awarded him an OBE for his key role in hunting down the Portland spy ring, July 1961.

Gordon Lonsdale in his White House flat. Note the radio used to receive messages from Moscow, as well as the Chinese scroll on the wall, which had a hidden cavity in the bottom cylinder. MI5 secretly searched the flat several times.

Spying in suburbia: Peter and Helen Kroger's modest bungalow at 45 Cranley Drive in Ruislip, with their car parked in the driveway.

Arthur 'Bill' Bonsall, Head of 'Z' Division and later Director of GCHQ: the Security Service point of contact to exploit GCHQ's expertise in breaking Lonsdale's codes and in trying to intercept the Krogers' KGB radio messages.

A dramatic rendezvous point for spies: the Old Vic theatre as it looked in 1960–61, when Lonsdale regularly met Houghton and Gee outside the main entrance.

Michał Goleniewski, the top-ranking Polish intelligence officer code-named 'Sniper', whose defection in Berlin on 4 January 1961 triggered the urgent arrest of the Portland spy ring three days later.

Hopefully disguised by his beret, Superintendent George G. Smith of Special Branch waits at Waterloo station in London for Houghton and Gee on 7 January 1960 (newspaper reconstruction photo).

Four stills from a 1963 US Defense Department film showing a reconstruction by actors of the 7 January arrests: Lonsdale arrives at Waterloo; Lonsdale greets Houghton and Gee outside the Old Vic; the arrests; and Lonsdale bundled into an MI5 car by Superintendent Smith (note Gee's shopping basket containing Portland secrets). The film was made with British cooperation at the original locations to warn the armed forces and civil servants about the dangers of Russian espionage.

Covert trapdoor in the floor of the Cohens' Cranley Drive kitchen.

Expertly concealed hiding place for the Cohens' radio equipment: it was only discovered by MI5 in the cellar of the bungalow several days after the arrests.

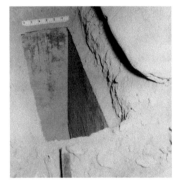

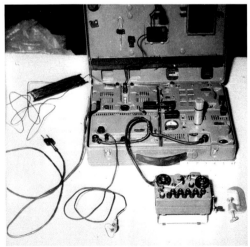

Ruislip calling Moscow: the Cohens' state-of-the-art 'flash transmission' radio equipment. GCHQ believed it was less than a year old and the Cohens had only used it rarely.

Man and master: Alan Belmont, head of the FBI Domestic Intelligence Division (left), with FBI Director J. Edgar Hoover.

The FBI's gigantic fingerprint centre in Washington DC in the late 1950s. The Bureau circulated the Cohens' fingerprints to police and intelligence agencies around the world in February 1958.

Microdots found in
Lona Cohen's handbag
when arrested.

Contents of one microdot: a letter
to Molody in London from his
family in Moscow.

A teenage Konon Molody in
California in around 1936: this
snapshot was given to the FBI by
his relatives in the USA.

An adolescent Molody
(second from right)
with his aunt Tatiana
(second from left) in
California.

Molody's Brussels mistress, Denise Peypers, photographed by her Russian lover inside her antiques shop.

Molody in Moscow with his wife, Galina, and children, Trofim and Elisabeta, during one of his visits home from London.

A spy for a spy: a Soviet photo of the moment Konon Molody was swapped for Greville Wynne at dawn on 22 April 1964 in Berlin.

Morris and Lona Cohen in celebratory mood on their
flight to Poland en route to Moscow on 24 October 1969,
after their spy swap.

Protected by local police from gawping neighbours,
Ethel Gee returns to her house on the Isle of Portland
after her release from prison on 12 May 1970.

Launch of Britain's first nuclear submarine, *Dreadnought*, on Trafalgar Day, 21 October 1960. By this date Houghton and Gee had already provided Molody with top-secret details of its world-beating sonar.

Type 2001 sonar in use in a British nuclear submarine. Secrets of its design helped the Russians develop their own version, which wags in the Royal Navy dubbed '2001sky'.

Molody in the Red Army (right). He rose from the ranks to become an officer in an artillery brigade, and fought at the second battle of Smolensk in 1943. 'After what I have experienced in the War, I have nothing to be afraid of', he told his KGB recruiter.

Spies in the post: the heroic stature of Molody and the Cohens in Russian eyes, commemorated in postage stamps in 1990 and 1998.

Pavel Gromushkin, the KGB's legendary forger, photographed in the SVR Press Bureau in Moscow. He forged important documents for the Cohens as illegals. Note his chestful of medals and the portraits on the wall behind of Russian espionage heroes – here displaying Morris Cohen (left) and Anthony Blunt.

The long view of intelligence history: SVR Director Sergey Naryshkin (left) chats with the grandson of the founder of the first KGB in 1917, Felix Dzerzhinsky, at the launch of a special exhibition of photos of 'golden' KGB spies, Moscow 2017.

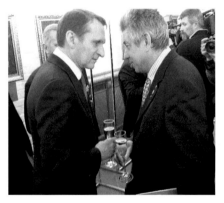

Molody's daughter, Elisabeta, at her father's grave in Moscow's Donskoy Monastery Cemetery.

spy'. As for the assistance Costello and his wife were giving the Soviets, the evidence pointed 'to an illegal support role for one or both
of them', with the professor 'well placed for talent-spotting among
students of Russian and intending visitors to Russia'. In December
MI5 warned the Security Department of the Foreign Office that
'both Costello and his wife are agents of the Russian Intelligence
Service. This information is completely firm. We do not know when
they were recruited by the RIS, but this must have been some time
ago, possibly when they were in Moscow.' The same month, Costello
was again tailed in London to a brief meeting with the same KGB
agent and seen to hand him a 'light-brown document case'. MI5
had yet further evidence that Costello was a KGB agent, but it was
impossible to interrogate him as hoped and planned because in
February 1964 the professor died suddenly. When he confessed to
being a Soviet agent later that year, Anthony Blunt named Costello
as a possible KGB spy (he was a Cambridge contemporary), and
the KGB defector Vasili Mitrokhin revealed twenty-eight years later
that Costello was described in a 1953 file in the Soviet intelligence
archives as 'a valuable agent' of the Paris *rezidentura*.[47]

It is now clear that both Paddy and Bella Costello were KGB
agents. Bella Costello knew about and was closely involved with
helping the KGB obtain 'dead double' passports and other support
work for illegals. Deprived by 1954 of any access to sensitive or
classified information because of (well-founded, as it now proves)
concerns about him being a security risk, Paddy Costello issued
Morris and Lona Cohen in Paris with their new identities as Peter
and Helen Kroger. Without his help, the KGB would not have been
able to set up the Portland spy ring as they did.

In the months immediately after the Krogers were imprisoned,
Elwell continued to investigate their background. Their letters,
neighbours and acquaintances, however, could add little of substance to what the Security Service already knew. Across the Atlantic, the FBI had more success. They interviewed or re-interviewed
friends and associates of the spies, exchanged information with the
British and Canadian intelligence services, and accumulated gargantuan files on the Cohens crammed with thousands of documents
which still remain secret.[48] While incarcerated Lona and Morris
Cohen exchanged numerous letters between themselves – always, of
course, assiduously signing all their missives in the names of Helen

and Peter Kroger – on the buff-coloured and lined notelets issued to prisoners. They spoke of their various health problems, international affairs and fantasised about the food they wished they could eat in their prison canteens. The correspondence helped to maintain their morale and was testament to their strong personal bond.[49] One British friend from the antiquarian book trade who wrote to them was Winnie Myers, based in St John's Wood in north-west London. In one letter to her written at the start of 1968, Kroger described his life in prison as 'monastic', with the 'remarkably subdued' prisoners over the holiday break watching the film *Scrooge* on television while eating their Christmas pudding. The Krogers were allowed well-guarded meetings at a 'neutral prison' every three months, but because of Helen Kroger's health problems her husband often found the encounters troubling. He arrived on one occasion in 1968 'like a freshly scrubbed, excited kid', but when he first saw her 'hobbling with her back bent and struggling to get a smile to break through' his 'heart [fell] bump! to the floor'. The meeting ended with them 'embracing to keep our hearts from slipping away'.[50]

Any spy swap depends on the nations involved having pieces on the diplomatic chessboard to exchange. Throughout 1964, after Lonsdale returned to Moscow, the Russians possessed nothing to bargain with for the Krogers' release. This changed early the next year.

Gerald Brooke was a British lecturer in Russian. He visited Moscow in April 1965 at the head of a group of student teachers, where he and his wife visited a Russian couple, passing them a number of items given to him by 'an acquaintance' in Britain, including a photograph album. When the Brookes visited the couple a second time, the Russian couple had just opened the padded cover of the photo album and were extracting anti-Soviet pamphlets from it when the KGB burst into the flat and arrested the Britons. The hapless Brookes had fallen into what was almost certainly a carefully laid KGB trap. Brooke was seen in Britain as an innocent victim and his case was taken up not only by the media but also with characteristic vim by his local MP and future Prime Minister, Margaret Thatcher. MI5 discovered that Brooke, although not involved in intelligence work, was deeply religious and involved with a Russian émigré group which engaged in propaganda work against the USSR and clandestine operations behind the Iron Curtain. As early as May 1965, it was suggested that the Krogers might be swapped for Brooke, but the

Foreign Office rejected the idea peremptorily: in its view the Krogers were 'big fish caught in a serious espionage case' and were 'in no way comparable with Mr Brooke'.[51]

Brooke's trial took place in July 1965. Moscow accorded it maximum publicity, holding it in a public theatre before an audience of 600, in front of TV and film cameras. Brooke was sentenced to five years in prison. Within weeks, in his KGB-inspired letters to the Krogers, Lonsdale was referring to 'behind-the-scenes negotiations in Whitehall concerning your case in connection with the Gerald Brooke affair'.[52]

MI5 hardened British government opinion against any early exchange by pointing to the differences between Wynne and Brooke: the former was a UK espionage agent working for the British authorities to whom the government owed an obligation, whereas Brooke was merely a private citizen acting as an individual. The Americans also opposed any swap. Some in Washington wished eventually to have the Cohens extradited back to the USA to be tried for espionage. To the British, Brooke was a foolish young man. To the Soviets, Brooke was working for an émigré organisation dedicated to undermining the USSR and funded by the CIA, and so had at least dipped a toe into the murky waters of intelligence. Moscow planted stories in the British media to influence public opinion, underlining the tough conditions of Brooke's imprisonment and suggesting that the East German lawyer, Wolfgang Vogel, who had helped engineer the Powers–Abel exchange, was now seeking to negotiate a similar deal for Brooke and the Krogers. To help block any attempt to extradite the Krogers to America, the KGB continued their various manoeuvres to establish that the Krogers were Polish and no longer American citizens, claiming they had gained Polish citizenship while resident in the country in 1950–54: a claim strongly contested by MI5.[53] Tension was heightened by the Polish government demanding that their consul in Britain should have access to the couple. The UK government continued to make pleas for clemency for Brooke, but Moscow remained intransigent. Relations cooled even further in 1968 after the Soviet invasion of Czechoslovakia.[54]

What finally broke the deadlock in early 1969 was a cynical blackmail threat by the Soviet government to bring new charges of espionage against Brooke. The prisoner was approaching the end of his sentence and Moscow faced losing their valuable bargaining chip. Once again the English lecturer was entrapped by the KGB, but this time in his labour camp. Fellow prisoners persuaded Brooke to discuss potential British espionage in the Soviet Union and dutifully reported their conversations. Brooke faced the dire prospect of a potential new sentence of fifteen years. His health was deteriorating. If Brooke was imprisoned for longer the British would need to retaliate, souring British–Soviet relations, just when a renewed dialogue between West and East to limit the superpowers' strategic nuclear arsenals seemed possible. Resistance in London started to crumble, although Martin Furnival Jones (now promoted to Director General of MI5) was among those who opposed early release for the Krogers, together with the Home Office, led by James Callaghan. The two sides slowly edged towards a compromise. The exchange would not be simultaneous. There would be a three-month delay between Brooke's release and the Krogers being freed; and Brooke's release would be slightly sweetened for the British by the Soviets repatriating two British hippies imprisoned for drug-smuggling.[55]

Just before midday on 24 July 1969, Brooke's flight from Moscow landed at Heathrow airport. The Krogers were informed of the deal in prison. Exhilarated but impatient, they needed to while away three months until they departed on a flight to Warsaw, the destination chosen by the Russians to be consistent with their claim to Polish citizenship. Helen Kroger in particular, according to her husband, did not allow herself 'to react to the culmination of a dream until she experiences the actual conclusion and climbs aboard the Warsaw-bound plane'. A few weeks before Peter Kroger's release, some prisoners at Parkhurst laid on a farewell dinner in his honour. They had saved up their prison wages and the party feasted on fish, chicken and ham. Kroger's chair was decorated with a handmade Soviet flag. The guest of honour made a brief speech, which was surprisingly apolitical, and greeted with a round of applause.[56]

The Krogers were flown out of Britain on 24 October 1969. Peter was fifty-nine, Helen fifty-six. When they were reunited just before the flight, Helen greeted Peter with the words, 'Hello, sweetie pie.' Their departure was an enormous news story. The civil servant who headed the Foreign Office and played a crucial role in the negotiations, Denis Greenhill, wrote later that the photograph of 'a defiant Mrs Kroger' on the steps to the aeroplane 'made me wonder, for a few moments, whether we had let our hearts run away with our heads'.

The success of the Soviet blackmail underlined how little value the USSR placed on warm relations with the British government, and how powerful the KGB was in the Soviet political apparatus.[57] On arrival in Warsaw the Krogers were whisked away from the waiting paparazzi for a medical check-up, before being flown secretly to Moscow the next day, to pick up the threads of their relationship and start their lives in exile.

10

Houghton and Gee

After the trial, shocked by his fifteen-year sentence and facing up to the gloomy prospect of losing over a decade of his life behind bars, Houghton soon became disillusioned and querulous in jail. Resentful of any sign that he and Gee, as convicted spies, were treated in any way differently to other prisoners, he believed that they were subject to a campaign of unfair discrimination and 'deliberate harassment' by the prison authorities. In fact, behind the scenes the Prison Commission gave special permission for them to write to each other (unmarried couples were not normally allowed to correspond while in prison) at the behest of MI5, which hoped to glean scraps of useful intelligence from their intercepted letters. They were, however, not allowed to meet.

Houghton believed their correspondence was deliberately delayed, and was outraged that when he met another visitor early on in his prison term a warder was present throughout and took notes. After a few months at Wormwood Scrubs, Houghton was transferred to Winchester. Gee, meanwhile, was something of a model prisoner and moved to Styal open prison in Cheshire, where she became a librarian. Although her nerves were 'in a shocking state', boredom appeared to be her worst problem, and she complained that she spent all her time 'baking bones and making soup'.[58] It was too distant for her elderly and infirm mother to visit but, when it became evident she was dying, Gee was allowed out briefly to be at her bedside for an hour or so at the start of 1966. Although intermittently depressed, she was more sanguine than Houghton about her predicament. She wrote to him from Styal that 'this is the first time since I have been in the "Nick" that I have really known peace of mind from outside, and if this continues I can do till 1971 standing on my head!'.[59]

Elwell made his first visit to Harry Houghton in Winchester prison on 18 May 1961, as with Lonsdale and the Krogers under the cover name of Charles Elton. He asked if Houghton was prepared to give 'a full and detailed account of his dealings with the Polish and later Russian intelligence service from the time that he started his career of espionage in Warsaw', emphasising that the Home Office would take into account Houghton's cooperation when his sentence was reviewed. One reason why Elwell and MI5 were so certain that Houghton had been spying for the Russians for a decade before his arrest was that MI6 had confirmed through two additional agents in Polish intelligence – in addition to 'Sniper' – that someone on the staff of the British military attachés in Warsaw had written to the Polish authorities in 1951 offering secrets in exchange for money. Another was that MI5 had unearthed confirmation of Houghton staying overnight in London on numerous Saturdays every year from 1955 onwards, on four occasions with Gee pretending to be his wife. The prisoner responded that 'he was only too willing to help', and referred to the note he had given his barrister to hand to the judges hearing his appeal which summarised the kinds of intelligence sought by the Russians and what they were already receiving. Elwell said the only type of assistance 'that would really be of help' was a detailed confession. Houghton retorted by denying that he had ever spied while in Warsaw and repeated at tedious length the lies he had told in the Old Bailey witness box. The only admission he made was that Gee did not, in fact, know Lonsdale as Alex Johnson but as 'Gordon'.[60] Elwell met Gee briefly at Holloway six days later. Embittered, she refused to add anything to what she had said in court and was wholly uncooperative. 'She said that she had nothing to tell me,' wrote Elwell, 'and with trembling lips got up to go.'[61]

Elwell's enquiries built up a more accurate picture of Houghton's time in Warsaw: his black-marketing ('a fair amount . . . but not as much as some people'); and his love of parties ('lots of NAAFI cigarettes and whisky and wild, wild Polish women'). Elwell also conducted a sympathetic follow-up interview with Houghton's former wife, Peggy, in December 1961.

Elwell even visited Gee's house in Portland, where he met her elderly uncle. Gee's mother 'had become very bitter against the police', he told Elwell, 'to the extent that *Dixon of Dock Green* [an immensely popular if anodyne police series of the time] was not now allowed on the television'. Elwell found Gee's uncle to be 'a very decent, sensible sort of man, who does not seek to excuse his niece's behaviour though he does feel that the sentence was perhaps a bit harsh'.[62]

On 12 September 1962 another KGB spy in the Admiralty was arrested, John Vassall. Sentenced to eighteen years in prison for his treachery, this was yet another espionage depth charge for Macmillan. Its effects were heightened by a report in the *Daily Express* in early November, after the trial, that the Lonsdale network and Vassall were connected, and that MI5 had found a top-secret document during their search of the Krogers' bungalow indicating that there was a second spy burrowing into the Admiralty. Lord Carrington immediately summoned Hollis, who underlined that the newspaper article was 'very misleading', and the First Lord said he was going to press Macmillan to make a statement contradicting it. A few days later, as the political storm clouds gathered, the government set up a tribunal to inquire into the Vassall case. Elwell (who led the MI5 Vassall investigation) and his Security Service colleagues quickly guessed that the *Express* article was inspired in some way by Houghton and his note handed to the appeal judges referring to intelligence the KGB was receiving from other spies. Sensing potential trouble, Hollis demanded why his counter-espionage officers had not delved further into the note when Elwell had first interviewed Houghton in prison. Elwell countered by saying that if 'I had questioned [Houghton] about his note I should have fostered in him the illusion that he was helping the authorities' and encouraged his schemes to reduce his sentence or even obtain a pardon. Elwell considered that without first a full and frank confession by Houghton of his espionage, it was impossible to evaluate the reliability of the information in the note. Hollis met the Prime Minister on 21 November to explain the Houghton note. Macmillan commented that 'he did not see how the *Daily Express* could maintain that their article was based upon it'.[63]

Before the Vassall tribunal could start its hearings, the press heard about letters found in Vassall's possession from Tam Galbraith, the Civil Lord of the Admiralty, who was number two to Lord Carrington. Newspapers suggested that it was odd that a government minister would communicate by post with a junior official of his own department, and there was speculation of a possible homosexual relationship between them, and of Galbraith shielding Vassall. Macmillan was forced to widen the inquiry to cover the issue of whether the Admiralty should have known much earlier about a second spy, and sharpen it by appointing a judge to lead it, the formidable Lord Radcliffe. Hollis's instincts, as so often on threats to his Service's reputation, were sound. The inquiry soon started 'taking a great interest' in Houghton's note, and Houghton, Hollis and Elwell were all called to give evidence. The convicted spy clearly relished his moment in the tribunal's private spotlight on 9 January 1963. His testimony was, typically, unconvincing and unclear, but he delighted in painting MI5 as spurning his offer of help to the authorities. Elwell appeared the same day and defended the way he had interviewed Houghton with polite directness: 'It is never worth asking a liar for information you cannot check.' The tribunal queried why the Security Service decided not to investigate Houghton's note further. Hollis was unrepentant in his testimony, pointing out all the reasons why in his view Houghton's note was unreliable and did not provide in any way 'a genuine target' for a sensible counter-espionage investigation, and why MI5 had made the correct decision. Under cross-examination, however, Hollis admitted that perhaps Elwell's approach to Houghton in the interviews had been too robust, in effect telling Houghton he was a liar to his face. The inquiry report did not criticise MI5 in any way but, behind the scenes, Hollis was clearly stung by the tribunal's interrogation. It was decided to send another MI5 officer to parley with Houghton, and hold out the prospect of meetings with Gee as bait.[64]

This was Jim Patrick of D4, responsible for counter-espionage agent-running. His first meeting with Houghton was on 26 March 1963 in Winchester prison. The loneliness of prison life had soured to bitterness. Houghton complained repeatedly that he was not given a fair trial (largely, he said, because he had been refused permission to call twenty-seven witnesses in his defence and his attempts to help the authorities had been spurned), and he would publish his story after release from prison, confessing to Patrick that 'he would not hesitate to pad his book with fictitious stories in order to make it more sensational'. After quizzing the prisoner about his note on Russian intelligence targets that caused MI5 so much discomfort at the Vassall inquiry, Patrick concluded that the contents were 'no more than an intelligent guess on his [Houghton's] part'. Houghton made it clear that he would only talk frankly about his espionage if he and Gee were released. Houghton intermittently toyed with the

Security Service in the years that followed, hinting again to Patrick in April 1964 that he would be prepared to talk if he and Gee were freed – but he would do nothing without Gee's consent. At this 1964 meeting with MI5 he admitted for the first time that his career as a spy had 'covered a much longer period' than the court had been told at his trial and that the story he told in the witness box in his defence was 'all eyewash'.[65]

Gee's attitude remained obstructive. She wrote to Houghton in May 1964 that 'under no circumstances would I do any sort of deal with [the Security Service]'. The couple were finally allowed to meet in September 1964 and were covertly recorded by MI5. Houghton told Gee he had 'received a couple of visits from the enemy [the Security Service]', but 'we're not getting anywhere'. She cut in, according to the transcript, 'in a biting tone, to say "Harry, look, if ever I thought you told them—"'; he interrupted 'I shan't'; and Gee continued, 'don't speak to them in any way; I'd be back again doing another sentence'. This exchange suggested that Gee was concerned that if the full truth of her espionage emerged she faced an even longer term in prison. MI5's trawl in 1964 through Gee's contacts during her time at UDE shed some patchy light on her possible motives for agreeing to spy with Houghton other than love: she often appeared 'thoroughly bored' and regularly complained of Portland being a 'dead end'.[66]

Having always displayed an unattractive combination of entitlement and resentment, and, now in prison, also an attitude of competitive victimhood, Houghton was especially piqued by his treatment after the George Blake escape, when he was hustled up to Durham prison, placed in solitary confinement for twenty-eight days and as a 'special watch' prisoner was compelled to wear a distinctive uniform of different patterned materials. Gee was driven down to the security wing at Holloway and, as the authorities panicked, suffered similar harsh treatment. Houghton was later transferred to Maidstone prison in Kent, where there was a more relaxed atmosphere but he was (he claimed later) attacked several times by young thugs.[67]

In August 1969, with the release of the Krogers impending, the Home Secretary, James Callaghan, felt obliged to consider releasing Houghton and Gee early. Callaghan was frank in his note to the Prime Minister, Harold Wilson, that the deal to free the Krogers was 'political', but all prisoners including spies like Houghton and Gee serving sentences longer than eighteen months were eligible for early release on licence. The local review committee in both 1968 and 1969 reported favourably on Gee, but not Houghton. They commented that while Houghton's 'prison performance is regarded as marginally good, the committee are concerned by the *fact that there is no evidence of regret on his part; this is considered to be particularly important in view of the nature and seriousness of his offence*'. (Wilson underlined these words with blue ink.) Callaghan said he was ready to release Gee a week before the Krogers, but Houghton should stay in prison. In response the Prime Minister simply scrawled 'What??' on the note. Callaghan was overruled.[68] Both Houghton and Gee were freed on the same day, almost six months after the Krogers, on 12 May 1970. It was a bitter irony that, although the jail sentences of Houghton and Gee were the briefest of all the Portland spies, they languished the longest time in prison.

After being driven out of Maidstone prison, concealed under a blanket on the floor of a taxi to avoid the press, Houghton took a train to Poole in Dorset where he was greeted by a posse of journalists. On the day of his release, Houghton issued a statement saying that he was responsible 'for the tragic events in which [Ethel Gee] became involved and . . . she has never once reproached me . . . such love is hard to find'. On her release from Styal open prison in Cheshire, Ethel Gee arrived back nine hours later at her old house in Portland, empty now after the death of her mother and elderly relatives.

Part Six
Russian Versions of the Truth

11

Konon Molody in Moscow

On his return to Moscow in March 1964, Molody was greeted with jubilation but also suspicion. There was joy at his homecoming in the comfortable flat provided for the family by the KGB on the Frunzenskaya embankment, where many secret service personnel were housed. For the first time he was united with his family for a prolonged period: his long-suffering wife, Galina; daughter, Lisa; and six-year-old Trofim, who had only seen his father for brief periods each year until his imprisonment in 1961. No doubt there were family tensions, but years later in his memoirs of his father Trofim testified to his parents' 'really strange' marriage, and the love Molody bore his wife: 'maybe because he saw her rarely, or more likely because of me'. Willie Fisher (Rudolf Abel) and other KGB acquaintances and relatives called round for celebratory drinks, and he and Molody resumed their friendship. Their KGB dachas outside Moscow were close and Molody used to visit Fisher at his flat in Moscow to play chess and discuss espionage.[1]

Like any patriotic Soviet communist, but especially one on his return to Moscow after three years in British prisons, Molody would have feasted on the sights of the Russian capital. Moscow had been transformed in the decades of Stalin's dictatorship into an exhibition of Soviet power: a stately underground railway rumbled under the city and skyscrapers pierced the sky like tiered cakes. Wide boulevards lined with monumental architecture and tower blocks sliced through the pre-revolutionary maze of streets. In 1964, in front of the imposing hulk of the Lubyanka, in the centre of the traffic circle, stood the tall and solemn statue of Felix Dzerzhinsky, founder of the Cheka, the first ancestor of the KGB.[2] Molody would have found the interior of the KGB headquarters little changed from his previous visits during absences from Britain when he had filed detailed reports on his work as the illegal resident in London. Long corridors lined with strips of red carpet ran through the building, opening on to small rooms shared by several KGB officers, or grandiose ones for more senior personnel. As with MI5 and the FBI, there were typing pools for carefully vetted secretaries and, like those intelligence agencies, the KGB also acted as an unofficial matrimonial introduction service.[3]

Lurking beneath the smiles and handshakes of welcome at the Centre was suspicion. Molody, in the words of one former senior officer of the KGB, Mikhail Lyubimov, was 'greeted with great reserve back in Moscow'. The implosion of the Portland spy ring was bemoaned as a fiasco in the Lubyanka, where there were fierce arguments and political manoeuvring behind closed doors, and gloated over by its rivals. Ivan Serov, for example, was head of Russia's long-established and powerful competitor to the KGB, the military intelligence agency the GRU, and embittered after being demoted from the post of Chairman of the KGB by Khrushchev in 1958 because of his deserved reputation as a brutal thug. He pointed an accusing finger at his successor at the KGB, Alexander Shelepin, in a 1961 diary entry made soon after the Portland arrests:

He [Molody] was a good man. I was there when he was sent by the KGB [to London]. He worked well for several years. Shelepin's people probably made some mistakes. And that's why the English could find him [Molody]. With him they also arrested other people but the most important thing found was the high-speed radio transmission equipment which for us is really significant . . . Nothing has happened to Shelepin because he is in favour . . . The KGB told me they have recalled dozens of people because Lonsdale and others knew them.[4]

The final comment hints at the disarray caused to the Centre by the Portland arrests.

Many senior officials in Moscow blamed Molody personally for the debacle: he had been slack with his tradecraft and lured MI5 surveillance to the Krogers. Others demurred. Chief among these was the KGB officer in the London embassy, Vasili Dozhdalev, whose London address had been found in Houghton's cottage and who had fled soon after the arrests. Dozhdalev – open-hearted, charming and devoted to the KGB – was well regarded because of his experience in successfully running valuable agents (like George Blake in Berlin in 1955, and Houghton before the Centre decided to transfer him to Molody). Dozhdalev considered the Centre had handled the whole affair badly and were foolish to have shuffled responsibility for controlling Houghton from Dozhdalev to a valuable illegal, Molody, without sufficient reason. The danger for the illegal was too great: once he came under surveillance by a Western intelligence service with the resources at their disposal he was doomed, because no 'legend' was perfect.[5]

Molody was astute enough to tread with care when he was debriefed relentlessly in the months after his return from exile. The information he provided would have been fed into the KGB's continuing inquest into the demise of the Portland ring, one of its most valuable networks of illegals. Although in many ways Cold War Russia was sclerotic and corrupt, KGB post-mortems on espionage failures were usually unflinching. As part of his interrogations Molody would have learnt of Goleniewski's defection to America in January 1961 and the KGB investigation that followed into his treachery, as seen from the Lubyanka. Molody shared the KGB's shock and outrage at Goleniewski's double-dealing and that he had been betrayed by a trusted senior officer of a satellite intelligence service. As he wrote a year later in his memoirs: 'Uneasy rest his head!'[6]

Goleniewski's defection proved a watershed for the KGB. An internal commission had been set up to study the case and lessons were learnt. One of the most important was the need for the Centre in future to be more circumspect in its dealings with the intelligence services of East European satellite countries, including the Stasi (which, for example, around this time, it seems, assassinated a person whom the KGB had targeted for recruitment), especially in the secrets it shared with them. Goleniewski had been trusted completely by the Russians and betrayed them. In future intelligence officers of the satellite states would be regarded with more suspicion. As for the Polish intelligence service, the KGB never fully trusted it again with any matter of great importance: as the Goleniewski case had demonstrated, its tradecraft and administration were sloppy, and because of the visceral hostility of many Poles to Russia, the Centre considered its officers were more likely to defect.[7]

After his return to Moscow Molody was nothing if not resilient. He held his nerve during the lengthy and no doubt ostensibly amicable interrogations and slowly the tide of disapproval and suspicion receded. He was also lucky. The KGB either did not learn about Molody's overtures to the British, or the spy was able to convince the Centre that they were an elaborate ploy to double-cross MI5. Also, through the bookseller friend of the Krogers, Freddie Snelling, a British publisher approached Molody in late 1964 and broached the idea of him writing his memoirs. There was a predictably sharp debate in Moscow about whether to allow the book to proceed. The final decision was elevated to the Soviet Politburo. Supporters argued that the book would burnish the KGB's reputation, but what seemed to have ultimately swayed doubters was a statement by the KGB Chairman that the publishing advance would buy seventy-five Volga cars for his organisation. Leonid Brezhnev, the powerful Communist Party First Secretary, agreed. In the months that followed, the memoirs were 'ghosted' by a man who had no peer for manipulating the truth, Kim Philby, and who had defected to Moscow the previous year. Each day, according to Philby's wife, he would travel to his KGB office to work on the book and return home each night 'completely drunk'. When asked a couple of years later if he had written the Lonsdale memoirs, the master spy admitted with laconic understatement, 'Gordon [Lonsdale] is a very gifted young man, but he's not a writer. I revised the manuscript.'[8]

The drafts were understandably scrutinised by the KGB and the book contains many lies: Molody clung throughout to his discredited 'legend' that he was Gordon Lonsdale from Canada, he had met the Krogers by accident in Paris and they were innocent dupes. He also falsely suggested that his wife was Polish and he had first met Rudolf Abel while fighting as a partisan on the Eastern Front. Throughout the memoirs, *Spy*, published in the UK and the USA in 1965, Molody and Philby seeded sly digs at the Western democracies and propaganda for Soviet communism. According to Molody's son, his father was not paid a single rouble for his contribution to the book. One avid reader was Charles Elwell of MI5. He was amused (and probably a little irritated) to find himself portrayed in the memoirs as an underpaid and ingratiating fool when interviewing Molody in prison, but pleased to spot some significant inconsistencies between the life of the real Konon Molody and that of his 'dead double', Gordon Lonsdale, as narrated in the book.[9]

Having rehabilitated himself in the eyes of the KGB, Molody was encouraged to give regular lectures about illegal work in the West at the First Chief Directorate training school. His sessions were popular and attended by young and ambitious KGB officers like Mikhail Lyubimov and Oleg Gordievsky. A flavour of those talks was given in extracts from interviews with Molody, published after his death, in a Soviet newspaper in which he described his callous but professional approach to recruiting a spy: a good prospect was one who held a middle-ranking but not key position with access to information, did not aspire to rise higher, had some form of chip on his shoulder about being a failure, drank copiously and had 'a weakness for the fair sex' (both being 'expensive habits'), and was critical of his own government. At the beginning of recruitment, the KGB resident 'must make it look as though he is not recruiting . . . at all, but is simply buying information . . . Once he has his claws into the agent, there's no getting away.' Lyubimov remembers Molody as talkative and charming with an excellent sense of humour, although inclined to exaggerate and fantasise. Chief among these fantasies was the lie Molody peddled with some success in Moscow that he was a successful businessman while in London. He showed Gordievsky with pride a photograph of the electronic locking device for a car which won a gold medal in 1960 at the International Inventors Exhibition in Brussels, and claimed he even made a substantial profit for the KGB. In reality none of the companies Molody was involved with was successful, and one even went bankrupt.[10]

Molody benefited from the appointment of a new Chairman of the KGB, Yuri Andropov, in 1967. A party *apparatchik* responsible for the brutal suppression of the Hungarian uprising in 1956, Andropov learnt from that experience the need for flexibility. Not only were Eastern European satellites to be treated with greater consideration, but Andropov wished to soften and enhance the image of the KGB within the USSR as a whole. And so in 1967, aware of the outstanding success of the James Bond film franchise in projecting a glamorous and successful image of Western intelligence, the idea of producing the first and most famous of all USSR spy films, *Dead Season*, was conceived.[11] Directed by Savva Kulish, and starring the Lithuanian actor Donatas Banionis as a KGB spy with the cover name of Lonsfield who sold jukeboxes, the plot has echoes of Molody's espionage in the UK, where one of his tasks was to penetrate the British chemical and biological warfare research centre at Porton Down. In the film, the Lonsfield character must track down a Nazi war criminal who has created a deadly chemical gas before he can unleash his weapon of mass destruction. The film was released in December 1968. Molody was consulted during production, but Russian sources suggest it was not a happy experience for him, although it helped burnish his status within Russia as a celebrated spy.[12]

Molody's life, however, was far from untroubled. His time in the West clearly influenced his views and George Blake (who associated briefly with him in Moscow) confirmed that he found it difficult to adjust back to 'the restricted life of a KGB officer and was impatient of many aspects of Soviet reality'. He was especially critical of 'the inefficient and often incompetent way Soviet industrial enterprises were run and international trade conducted . . . and found himself relegated to a position of relatively minor importance'. There are stories that Molody became an alcoholic, although these are – perhaps predictably – contradicted in Russian sources. Other unconfirmed reports suggest Molody suffered health problems around 1969–70, leading to KGB doctors giving him injections which caused severe headaches.[13]

On a Saturday in October 1970, Molody was on a mushroom-collecting expedition with his family and two friends near the town of Medyn, about 125 miles from Moscow. During a picnic, immediately after his second glass of vodka, he suffered a severe stroke. Aged only forty-eight, he collapsed and lost the power of speech. In a panic, they sped the stricken man to the nearest hospital and from there contacted Willie Fisher. The veteran agent in turn called the KGB, which despatched a vehicle to pick him up. Molody died in hospital a few days later. On a funeral bier in the KGB officers' club his body lay in state, surrounded by velvet cushions displaying his impressive collection of medals. Yuri Andropov, and other KGB dignitaries, visited to pay their respects.[14]

A clear indication of Molody's importance was the decision taken soon afterwards to include him as the only post-Second World War illegal in the Memory Room of the First Chief Directorate at the KGB's new headquarters at Yasenevo, on the outskirts of Moscow, when the service moved there in 1972. Molody took his place among the stars and names incised on the walls of the circular room alongside other outstanding officers and agents of the service.

In 1976, a monument costing 2,000 roubles was erected on his grave in Moscow's Donskoy cemetery next to the tomb of his friend and fellow illegal, Willie Fisher. Molody's wife sought consolation in drink in the years after his untimely death and was treated several times for alcoholism.[15]

12

Konon Molody – the spy

I

Twenty-one years after Molody's death, when in December 1991 the Deputy Director General of MI5 flew into Moscow's Sheremetyevo airport for two days of extraordinary meetings with the top brass of the KGB, she did not expect to be greeted by a welcoming committee of bemused men, the leader clutching a bunch of red roses. Stella Rimington had risen to the top of the Security Service to become its first ever female Deputy (and in 1992 first Director General) through a combination of toughness, ambition and luck, and she wished to be greeted as an intelligence professional first and foremost, and as a woman second. The floral welcome did not bode well for the meeting. She was in Russia at the invitation of its leader, Boris Yeltsin, to help advise the reshaped KGB on how to operate in a more democratic and open society. That same month the Russian Federation split the former and mighty KGB into two: broadly, the SVR took over the work of foreign intelligence formerly led by the First Chief Directorate, and the FSB the domestic espionage role of the KGB. After staying overnight in the British embassy, Rimington was whisked the next day through snowbound streets to a meeting in a cavernous room in the neo-classical hulk of the former KGB headquarters at the Lubyanka to meet the most senior officials of Russia's new espionage establishment. The British talk round the conference table was all of oversight and democracy, but the habits of decades were impossible to throw off. After the meeting Rimington observed ruefully that when she explored Moscow, she was tailed everywhere by a KGB surveillance team.[16]

The efforts to democratise Russian intelligence faltered in the years that followed, despite some covert but concrete assistance from British intelligence, including the provision of surveillance equipment to a Russian government department which appears to have ended up with the nascent FSB.[17] That extraordinary December 1991 meeting presaged the start of a heady but brief period of relative openness about the history of Soviet intelligence. For a few years limited access was allowed for the first time to some of the KGB archives covering the decades up to and during the early Cold War, including those on the Cohens and on Houghton and Gee. Intriguingly, the files on Molody stayed closed. Books and articles by former KGB officers and Western writers were published about all these spies and Soviet espionage in general, based on Russian sources that were tightly shut only a few years before. Some were printed only in Russian. The 1990s also saw the publication of a treasure trove of information from Russian defectors to the West, like Oleg Gordievsky and Vasili Mitrokhin. Russia and its intelligence services experienced this period as one of profound dislocation. Fearful of being forced to expose details of the thousands of people who had in various ways cooperated with the state, KGB officers burnt file after file and nursed a bitter sense of loss and hurt. As one later recollected: 'Those of us who had grown up as proud citizens of a superpower (even if aware of its many flaws), whose heads had been filled with stories of the advances and glories of the Soviet regime, spent the 1990s living in a land we barely recognised.'[18]

With the rise of Vladimir Putin, and the reassertion of influence by those within Russian intelligence who cleaved to the old ways and disapproved of the revelation of sensitive KGB material, the curtain of secrecy was slowly drawn shut. The KGB archives were once again sealed.

The possible problems with the documents about intelligence operations released into Western archives (including those in the UK National Archives about the Portland spy ring) are different. Clearly, these files can only be selected from those which have already been saved from destruction, and in turn can be pruned or redacted for a variety of reasons – some perfectly sound and some potentially self-serving. Historians have even been inveigled into writing an 'official' history of certain events, deliberately omitting important information and projecting a misleading narrative, to help keep secrets thought to be in the interests of national security. It is important always to bear in mind, as a leading British intelligence historian of the Cold War has warned, not to view the selected materials in government archives as an 'analogue of reality'.[19] Although

many MI5 files on the Portland case have been declassified, some
have not – most notably, the MI5 Kroger files from 7 February 1961
onwards.[20] No MI6 or GCHQ Portland files have been released,
and none is likely to be in the near future. A number of individual
documents in the published Security Service Portland files have also
either been withheld completely or been redacted. It is impossible
for a researcher to find out exactly why. MI5 document 'weeders'
must follow strict guidelines: for example, to protect the identity
of individuals who agreed to help the Security Service on the basis
of continuing anonymity; or only to release material involving a
friendly intelligence service with their express permission. (The FBI
and RCMP in this respect agreed to the publication of many of their
documents in MI5's Portland files; the CIA agreed to none.)

The historian must therefore question the released Security Service files, and supplement them through other evidence. With the
Portland case, this is fortunately quite abundant: for example, the
detailed and independent Romer Inquiry, other government papers,
memoirs and secondary sources, and even a few living witnesses.
The (now declassified) Romer Report contained some mild but hitherto secret criticism of MI5, and the Security Service have released
the papers which form the background to that criticism. There is no
evidence that the MI5 Portland files released in the UK have been
selected or manipulated with the deliberate aim of presenting a
particular view of the Security Service or its history. They certainly
do not tell the whole story – perhaps the most significant omission
being papers documenting the detail of GCHQ's crucial role in the
investigation. But for the first time they provide a firm archival foundation on which a historian can construct a fairly complete account.

The same emphatically cannot be said in Russia of either intelligence history in general, or of the Portland spy ring in particular.
The history of Soviet intelligence has become a servant of Putin's
Russian nationalist state, whose icons are not only religious saints
but spies. Secrets, however, once released, cannot be re-bottled.
Modern Russia and Russians cannot be controlled as comprehensively as their Cold War predecessors. These factors, together with
the continuing focus under President Putin on lauding the work of
Russian espionage, have ensured a trickle from the year 2000 (when
Putin was first elected President) onwards of new books, articles
and documentaries in Russian on the Cold War KGB. Very few have

been translated into English or other languages. To some extent this is understandable. The Russian Federation has no school of independent and respected historians of its national security and intelligence communities as have developed since the 1970s in the West. Authors tend to be former KGB officers or journalists close to and trusted by the SVR. These Russian sources must be approached with scepticism, like any other evidence. A few are simply risible in parts, repeating for example the hilarious *canards* that Konon Molody was knighted by the Queen, or became a billionaire businessman while in England, so rich that he was able to buy a luxury yacht from which, while sailing in the middle of the Atlantic, he was picked up by a Soviet submarine to be transported home to Russia.[21] Many, however, especially those written by former KGB officers with direct knowledge of events, are of value and merit study. Taken together with other material now available outside Russia, they help paint the most complete picture yet of the lives after release from prison of Molody, the Cohens, Houghton and Gee and, more importantly, of their espionage: what the Portland spies did, how they did it, and the damage they caused to the West.

First, Konon Trofimovich Molody.

II

Russian sources confirm the excellent sleuthing work the FBI carried out in 1961 to identify Gordon Lonsdale as Konon Molody. Although for the most part they merely confirm the story of his early years in Russia and California, they do add fascinating detail about his life back in the USSR when he returned from exile in America. He joined the Red Army in October 1940 and served from June 1941 as an intelligence scout with an artillery brigade, taking part in the 1943 Battle of Smolensk and travelling across Eastern Europe as German forces retreated. According to his army leaving certificate, he had risen from the ranks to become assistant chief of staff to the head of intelligence within the artillery brigade when he was demobbed in 1946, and he was awarded a medal specifically for his intelligence work. Some of this was undoubtedly perilous and required great courage, such as occasional forays when Soviet soldiers disguised in captured German uniforms penetrated behind enemy lines to kidnap soldiers as sources of intelligence. Many of Molody's school friends perished in the conflict. In 1951, when as part of KGB recruitment he was asked if he was prepared to face the dangers of becoming an illegal, he responded, 'After what I experienced in the war, there's nothing I'm afraid of.'[22]

After the war, from August 1946 Molody studied Law and Chinese at the Moscow University of Foreign Trade. It was here that his venery first showed itself: a fellow male student recalled how Molody one evening engineered a triple date with three married women, which culminated in the three students passing the night in separate rooms, each with one of the women. He also displayed a penchant for student theatre and amateur acting – an interest he shared, coincidentally, with the young Morris Cohen, and which no doubt stood them in good stead in later years when acting in the real-life theatre of KGB illegals.[23] It was in his third year at university that Molody was first talent-spotted by the KGB because of his fluent English. He was watched attentively and thought to be fine raw material for an illegal because of his self-confidence, boundless optimism, quick wits and hard work (he co-authored a textbook on Chinese in his spare time, for example). Around 1951 he was recruited to the KGB by Vitali Pavlov, who at that time headed the section of the service at Moscow Centre dealing with illegals in the UK and USA. Pavlov himself was a grizzled KGB veteran who specialised in North America, having been the KGB *rezident* in Ottawa in 1942 until he was expelled in 1945 in the tumultuous aftermath of the defection of Igor Gouzenko. Pavlov was highly professional, decisive and immensely self-confident, giving the impression to underlings that he knew everything, and was nicknamed by fellow KGB officers 'Cross Eyed' because of his slight strabismus. Back at the Centre, it was Pavlov who controlled Willie Fisher in the USA and who had oversight of Molody while he operated as an illegal in Britain.[24]

Molody underwent rigorous training in the forerunner of the KGB 'Red Banner' training academy, code-named School 101, shrouded in dense woods fifty miles north of Moscow. His training would not have differed greatly from that given to Oleg Gordievsky eleven years later when recruited to the KGB's Directorate S for illegals. Some of the most vital training was how to detect and avoid surveillance, known as *proverka* or 'dry-cleaning' in KGB parlance.

The trainees underwent various tests to learn how to spot if they were being tailed, shake off the surveillance without arousing undue suspicion, and arrive at an agreed location 'dry-cleaned' to meet a contact. Other training covered crucial KGB tradecraft: how to make a 'brush contact', passing an object or message to another person without being detected; ciphers and codes; secret writing; and microdots, photography and disguise. There were also the ineluctable and tedious classes in the ideology of Marxism-Leninism.[25] Molody impressed his teachers, one of whom took particular note of his knowledge of English and Chinese and his sense of humour. It is likely that he was given his first official false identity at this time: some documents in KGB archives about Molody are in the name of Konstantin Trofimovich Perfiliev.[26]

What Molody did between 1951 and 1954 is still cloaked in some mystery. Memoirs co-written by Molody's son and published in Russia in 1998 repeat and embellish the story told in Molody's own 1965 memoirs that he worked with Willie Fisher in America. According to this version of history, at the end of 1951 the Centre decided that Molody should be infiltrated into the USA to work with Fisher and complete his training as an illegal. The new arrival adopted with a certain amount of difficulty the 'legend' of being a commercial traveller by the end of the summer of 1952 and he had some responsibility for Fisher's communications and courier system, carrying out low-level espionage assignments, investigating former Nazis living in America and trying to obtain information about the military intentions of the USA and the activities of the CIA and FBI. In early 1954, Fisher told Molody he was being sent to England to function independently as an illegal.[27] The vast majority of Russian sources, including the SVR official history, are, however, silent on what Molody did between 1951 and 1954. It is quite possible that Molody spent this time in intensive training in Russia preparing for his first overseas assignment and never worked with Fisher in America for any length of time, but knew the older man from encounters in the Russian capital.[28]

The plan for Molody (code-named 'Ben' by the KGB) to establish an illegal *rezidentura* in England was in gestation for several months from the end of 1953 in the gloomy but spotless corridors of the Lubyanka, under the supervision of the tall and athletic former head of the Illegals Directorate, Alexander Mikhailovich 'Sasha' Korotkov.

By 19 March 1954 he was acting head of foreign intelligence in the KGB, and a document in the KGB archives shows that on that day Korotkov had written to the Chairman of the KGB, Ivan Serov. Serov was a diminutive but brutal man, a zealous Russian nationalist who played a major role in organising the infamous Katyn massacre in May 1940 of numerous Polish intellectuals, and in 1944 hunted down the remnants of the Polish Home Army with remorseless cruelty. In his report Korotkov informed Serov that the plans for 'Ben's illegal residency in England had been developed, with Morris and Lona Cohen to act as his radio operators once he was established, and he sought Serov's approval to go ahead. Serov assented a few days later.[29]

By November the Centre's preparations were complete. According to Vitali Pavlov, these had included his detailed briefing of Molody about life in Canada (Pavlov had spent three years in Ottawa as KGB resident). Molody was to travel to England by way of Canada.[30] Canada's passports had been a particular target for abuse by foreign intelligence agencies for many years. It was adjacent to America and, as a neighbour, presented numerous and convenient points to enter the USA. Abuse was facilitated by lax administration. Unlike the Communist Party of the USA, its Canadian brother had not been emasculated by draconian national security legislation introduced with the onset of the Cold War, and it gave the KGB considerable assistance throughout the 1950s. One example was how Molody entered Canada. A member of the Central Committee of the Canadian Communist Party had persuaded a party stalwart to give him his passport in 1953 'for Party use' when he learnt it had never been used for foreign travel. In fact, the Central Committee member passed it through a friendly intermediary to the KGB in Ottawa. The Centre replaced the original photograph in the passport with one of Molody, and it was brandishing this document that he entered Canada in the late autumn of 1954. Molody travelled by bus to Seattle and from there to Vancouver, using the identity of that Canadian communist 'live double'. As the RCMP and MI5 had learnt in 1960, with the aid of Canadian communists in Vancouver, Molody then applied for a birth certificate and driving licence in the name of his chosen 'dead double', Gordon Arnold Lonsdale. These two documents were the sole basis for Molody's application for a new Canadian passport in Lonsdale's name. To make this application, the new illegal needed help. Moscow Centre made a special, and no doubt quite painful, arrangement to ensure that Molody was secure when he linked up with the person who was to assist him. Molody later recalled how, before leaving for Canada, a dentist drilled several holes in his teeth. In Vancouver, the spy was instructed to visit a particular dentist who assisted the KGB, identify himself by quoting a famous line from a poem by Heinrich Heine ('Ich weiss nicht, was soll es bedeuten' – 'I don't know what it can mean') – then open his mouth and display the holes. Reassured that he was dealing with the correct agent, the dentist then filled the holes and helped him file an application for his passport in the name of Gordon Lonsdale, which was issued to him in Toronto in January 1955.[31]

Once in London, Molody stayed at the Royal Overseas League, where – according to his (probably misleading) notes published twenty-eight years after his death – he had a brief affair with one of the women administrators who helped introduce him to the capital: 'I am not a prude, just a normal person. You need to be either a fanatical monk or impotent to believe that a single "illegal" spy can live without a woman.' While working for the KGB in London over the coming years, it seems Molody pursued various women but, in keeping with his desire to boast, he vastly exaggerated the number of his conquests. His sexist and cynical attitude towards women was typical of the KGB of his time: there was a ban on the recruitment of female KGB officers for operational work which lasted until the late 1980s.

During his first days in London, he checked a telephone kiosk behind the Savoy Hotel regularly and when he finally found a map pin in the specified location under the shelf holding the directories, the newly arrived illegal hurried to a dead letter box where he discovered a message ordering him to visit Paris to meet 'Jean', a senior officer in the KGB. Having given the required signal of unusual complexity even for the KGB (standing at noon in the Louvre in front of Leonardo's *St John the Baptist* and scratching behind his left ear with a bandaged finger), Molody met 'Jean' on the Champs-Élysées and they talked in his car. The senior officer passed on a message that Molody's family were fine, and details of various contacts the illegal should meet back in Britain. Molody gave 'Jean' a letter for his wife, Galina, back in Moscow.[32]

According to one Russian source, possibly on the same trip, and also in Paris, Molody was instructed to meet the Cohens for the first time since they had arrived in England and rented a house in south-east London. Although the Cohens had encountered 'Ben' previously in Moscow, they had not been informed they were to work together in England. Following an instruction to travel to Paris to meet their new controller, Morris and Lona Cohen duly appeared at 5 p.m. at the Pyramides Métro station. There they were stunned to discover Molody loitering at the meeting place with the required copy of *Life* magazine in his left hand, and even more amazed when Cohen approached and employed the recognition phrase, 'I think we met in Warsaw last May?' and Molody replied with the correct response, 'No, it was in Rome.' Despite Cohen's protestations to stay calm, ebullient as ever, Lona Cohen assaulted Molody with a battery of questions about why they had not been told before that Molody (whose real identity they learnt only many years later) would join them in London to work in some as yet unspecified role. The three returned separately to England.[33]

After the Paris meeting in early 1955, for the KGB the way was now clear for the three illegals to establish themselves in London. Before learning details of their tradecraft and how they embellished their cover 'legends', it is helpful to leap forward in time to follow the lives of the Cohens after their spy swap in 1969.

13

The Cohens in Moscow

When the Cohens' flight from Warsaw landed on 25 October 1969 at Sheremetyevo airport in Moscow, their welcoming party from the KGB was deliberately unobtrusive. One man was familiar, Alexander Koreshkov, their KGB liaison officer while they had been based in Moscow in the early 1950s and their immediate controller at the Centre while they were based in London; the other was unknown to them – tall and elegant and almost two decades younger. There was a flurry of laughter when the younger man introduced himself in English as Lona Cohen's long-lost Polish cousin, 'Maria Petka': he had been writing to her in prison for several years as part of the Centre's stratagem to fabricate a believable Polish background for the couple. From the airport the Cohens were whisked away to a KGB safe house to be greeted by old partners in espionage: Willie Fisher, Molody and KGB officers from their time in New York.[34]

A month later, at a secluded KGB dacha, there was a dinner in honour of the Cohens. The guest list included all the KGB's top brass: Yuri Andropov presented the couple with the Order of the Red Banner for 'the successful accomplishment of special missions for the State Security Committee under difficult conditions in capitalist countries'; and Alexander Sakharovsky, the head of the First Chief Directorate, and Anatoly Lazarev, the head of the Illegals Directorate, also attended. Five thousand roubles were expended refurbishing a large apartment for the Cohens in Patriarch's Ponds in Moscow's Arbat district, only six blocks from the American embassy. The same KGB cast list attended a warming party there in April 1970. By Soviet standards it was luxurious, with the furnishing far more refined than the vast majority of apartments in the Russian capital. Around the same time, despite some internal opposition, the Cohens were granted Soviet citizenship.[35]

Morris and Lona Cohen (but as Peter and Helen Kroger) maintained a correspondence with a number of British friends once back in Moscow, using a PO Box in Warsaw to maintain the pretence that they were living in Poland. One was Nora Doel, the wife of the bookshop manager, Frank Doel, who was to be the subject of the 1970 epistolary memoir, *84 Charing Cross Road* by Helene Hanff, later transformed into both a play and a film. In one of his first letters to Nora, Cohen confided that 'During the first weeks after our return, we were treading on the clouds . . . Then the reaction came – with a terrific shock! Each of us got upset most easily.' The Cohens were sent by the KGB on a prolonged holiday around the Soviet Union to make the acquaintance of their adopted country.[36]

The Centre carefully regulated the Cohens' lives. They wished to avoid contact between the Cohens and other Western defectors in Moscow, partly because the Lubyanka wished to maintain the fiction that they were Polish and living in Poland consistent with the narrative developed while they were in British prisons. One day in June 1971, for example, Morris Cohen happened to encounter George Blake while out shopping. They expressed 'genuine joy' on their meeting, according to a KGB file in the archive, and agreed to meet again. The Centre rather callously ordered both Blake and Cohen to invent reasons to cancel the arrangement and avoid future contact. A KGB memo records a bugged phone conversation when Cohen called Blake to give his concocted excuse, and Blake on cue responded with his. The ban was lifted decades later around 1990, when the KGB asked Blake to make visits to keep the ageing couple company.[37] Lona learnt a reasonable amount of Russian, but her husband never assimilated enough of the language to converse with native Muscovites. This language hurdle, coupled with the KGB prohibition on the Cohens revealing anything of their true backgrounds, limited their ability to make new friends and socialise. Most visitors were from the KGB, some of whom were embarrassed by Lona Cohen's loud voice and brash manner in Moscow restaurants, where she would suddenly burst into song.[38]

With their unparalleled experience in the West as illegals, Directorate S of the KGB soon put the Cohens to work training recruits. The Cohens had entered the pantheon of KGB royalty together with spies like Willie Fisher, and were spoken of in reverential tones. They were introduced to trainee illegals merely as 'Peter and Helen'. Jack Barsky, the German illegal schooled by the KGB who worked undercover in the USA until he severed links with his Soviet masters in 1988, has described how he was tutored by the Cohens on how to talk, eat and behave. On the basis of Barsky's experience, which started in the summer of 1976, for the first part of training the fledgling illegal visited the Cohens once a week at their Moscow apartment. Morris Cohen was then sixty-six, but he appeared a decade older with wispy grey hair, a jaundiced hue to his wrinkled face, and arthritic, bony hands. Barsky was struck, however, by Cohen's hearty laugh and steely handshake, and the eagerness with which both the Cohens seized the chance to talk in English. Barsky noted that they 'didn't drink, they didn't party, they weren't in it for the money: instead, they were true soldiers of the revolution'. The mentoring culminated in Barsky spending a fortnight alone in a two-bedroom flat with Morris Cohen in February 1977 for an intense period of language and tradecraft training. Cohen imposed an almost military discipline, insisting that the day began at 6 a.m., and admonishing the trainee if he made a linguistic error which might arouse suspicion when undercover in North America. One day, gazing out of the window, Barsky saw a neighbouring building being demolished, and he murmured out loud, 'I wonder if they will *ironball* this thing down.' On hearing this invented verb, Cohen screamed, 'Don't you ever do that again! If you are experimenting with a language you have not yet mastered, you will be *dead meat*!' before stomping into another room. Barsky was terrified.[39]

Cohen suffered a stroke in early 1982, when he was seventy-two, and in letters he sent to friends in Britain he described his excellent medical care in a KGB hospital (carefully disguised, of course). While he was recovering at a KGB sanatorium (again, carefully disguised) with its 'slopes, trees and swift-running stream', Lona 'skipped around like a five-year-old', fell over a stone and broke her wrist.[40] Increasingly frail and depressed, and forced to use two sticks to walk, Cohen increasingly spent time isolated with Lona in their Moscow apartment, their domestic needs met by a KGB housekeeper.[41] With the reforms of openness and perestroika introduced by the new Soviet leader, Mikhail Gorbachev, the USSR began to totter. Cohen remained a true-born if unworldly communist as his health deteriorated in his final years, even asking an influential journalist to put him in contact with the General Secretary of the Communist Party because he wished to pass on some thoughts on how to modernise socialism. Old certainties crumbled. The previously unthinkable began to occur, including criticism of the KGB and its often brutal repression of Soviet citizens. The Centre responded in 1989 with a public relations counter-offensive.

A press office was established and a 'documentary' shown on Soviet television called *The KGB Today*. In 1990 a series of postage stamps were issued featuring great KGB spies, including the Cohens and Molody. The centrepiece of the publicity campaign was to be the revelation of details about one of the KGB's most spectacular but largely untold successes: the theft of the secrets of America's atomic bomb. The Cohens were part of that story. At the end of 1989 a retired KGB colonel, Vladimir Tchikov, who worked in the newly created press office, was much to his own surprise allowed access to the hitherto top-secret KGB files on the Cohens at the KGB First Chief Directorate headquarters on the southern out-skirts of Moscow to write a 'history' of their intelligence work. His draft was deliberately manipulated by General Yuri Drozdov, the head of the Directorate for Illegals, to attempt to mislead the West and its intelligence agencies about the KGB's spying network in the Los Alamos atomic bomb project and to protect Russia's ciphers. Nonetheless, Tchikov's book, based on access to over 6,000 pages of documents in the seventeen boxes of KGB files, and never published in English, is a unique source on the Cohens' espionage.[42] Assistance was provided to British and American television producers to film documentaries. Determined to make one last effort for their Soviet controllers, the Cohens rallied and agreed to be interviewed to record their memoirs. The KGB never allowed direct access to the Cohens themselves: the interviews were conducted by Russians approved by the Centre.[43]

By now Lona Cohen had developed lung cancer as the result of a lifetime of chain-smoking and the couple spent extended periods in a KGB nursing home in Moscow. There they were visited by a long-standing British friend of the couple, Sheila Wheeler, the daughter of a bookseller whose family had loyally supported the Cohens throughout the UK spy trial and afterwards. Lona Cohen justified her life of deception by believing that by helping to ensure Russia as well as America possessed nuclear weapons, this brought about a balance of power. Wheeler asked Lona at one point whether she thought her life had been a success. 'Well,' she retorted, 'we've had forty years of world peace. That's the achievement of my life.'

In other interviews around the same time, the Cohens belatedly acknowledged 'certain shortcomings in the Soviet Union', and one Russian historian even heard Lona characterise the Soviet system as 'outright totalitarianism'.[44]

In her final months Lona Cohen was admitted to a KGB hospital. In these twilight weeks, drugged by palliatives, drifting in and out of consciousness, her thoughts darkened. She asked a Russian historian who visited her one day, 'Am I a traitor?' She had no doubt interrogated herself on the same issue endless times before, but with death stalking her hospital room it no doubt acquired a new poignancy. Cohen answered the question herself. 'I didn't kill anybody and I didn't destroy any American life. No American soldier died because of what I have done.' In her last days, after some delicate negotiations, with the KGB pulling diplomatic strings behind the scenes, Lona's sister was flown from America to Moscow. The siblings, divided by ideology and thousands of miles of geography, had not met since 1950. It was the first contact Lona Cohen had had with her family for over forty years. She was not in the hospital room when her sister arrived. Morris Cohen stood by the window, staring out at the bleak Moscow skyline, wreathed in December snow. Lona appeared in her wheelchair a few moments later and the two sisters clutched each other and wept. In the few remaining days of her life, the two sisters leafed through old photos of their childhood brought from America. Lona Cohen died on 23 December 1992, aged seventy-nine.[45]

Morris was devastated by his wife's death. He became even more isolated and lonely in his flat. He told one visitor that he did not feel Lona had died, she had merely gone out to buy some bread. Under the marital bed, her slippers remained in their usual place. Occasionally his eyes would flicker up to a grainy black and white photograph on the wall which he had always cherished. It depicted a modest, columned, two-storey villa on the outskirts of Barcelona in 1938. To Morris Cohen it symbolised one of the most sacred moments of his life: the day when he pledged allegiance to the KGB.[46]

14

The Cohens as spies

I

When in August 1937 Morris Cohen clambered at night over the Pyrenees from France and down into Spain to join the fight of the International Brigades against Franco (the border was officially closed to foreign volunteers), his youthful idealism was soon to be tested. It was that idealism that had inspired him only weeks before to join the battle against fascism. An enormous rally had been held in Madison Square Garden in New York on 19 July against Franco and his Nationalist forces. Attended by around 20,000 people, it seethed with the enthusiasm of a revivalist meeting. Speakers roused the crowd to a frenzy, culminating in a call for volunteers to step forward to join the cause. To cheers from the crowd, Cohen was one who rose and took the long walk to the podium. Within days Cohen was assigned to the Mackenzie-Papineau Battalion of volunteers (their battle flag was red and gold with the motto, 'Fascism Shall Be Destroyed') and, as the FBI discovered in their investigation, had been given a false passport in the name of Israel Altman. Soon after his arrival in Spain, as a sign of his communist commitment Cohen was elected a political officer for the battalion, tending to the men's practical needs and ensuring their ideological purity. After infantry training the buoyant volunteers were sent to Zaragoza and then to the village of Fuentes de Ebro, eight miles away, to attack a Nationalist position. It was a military catastrophe. On 14 October at least sixty of the volunteers were killed and 150 wounded. Among them was Cohen: a machine-gun burst had sliced across his thighs and stomach. He was to pass the next four months convalescing in hospitals. For the remainder of his life he walked with a slight limp and blamed his wounds for his inability to father children. His optimism and ideological commitment, however, never wavered. An old friend from New York stumbled across him in the hospital: 'A chunk of his thigh was gone. The blood had seeped through his bandages . . . and was still oozing. "How are you doing, Moishe?" He responded with a beatific look of pure sainthood. "Just *great!*"[47]

Cohen's loyalty was reported on favourably by Communist Party functionaries and in early 1938 he was summoned to a mysterious villa on the outskirts of Barcelona. Owned by a Spanish aristocrat and surrounded by high walls, this two-storey building housed a covert sabotage, infiltration and espionage training school under the direction of the KGB resident in Spain, General Alexander Orlov. A hardened communist and secret service officer from the time of the 1917 Revolution, Orlov recruited Cohen and trained him to work for Soviet intelligence. That building was to remain a cynosure for Cohen for the rest of his life: the photograph that always hung on a wall of his Moscow flat.

As 1938 progressed it was clear that Franco's forces were in the ascendant. The Loyalist cause, and with it the International Brigades, were doomed to defeat. Meanwhile, in Russia Stalin's paranoia and purges had infected the KGB and hundreds of intelligence officers were being murdered. In November 1938 Morris Cohen steamed back across the Atlantic with other disillusioned American volunteers, but guarding a secret. Before departing for New York, the KGB had passed him half of a broken comb, and told him that in America he should wait to be contacted by someone holding the other half.

The following year, when Cohen was working at the Soviet Pavilion at the New York World Fair, a student at the Massachusetts Institute of Technology joined his table in the cafeteria. Elegantly dressed, slim and with a round face, he spoke excellent English and introduced himself as 'Sam'. After some desultory conversation, the stranger extracted from his pocket half of a broken comb which matched perfectly the half held by Cohen. 'Sam' was the cover name for Semyon Semyonov, an academically gifted Russian Jew who had studied electrical engineering in Moscow, where he had been recruited in 1937 by the KGB's new science section. The Centre had promptly sent Semyonov to MIT for graduate studies. By the end of 1939 Cohen was installed in New York at the Soviet company Amtorg, working in its restaurant. The following year, Semyonov joined Cohen at Amtorg under the cover of being an engineer. While studying at MIT, 'Sam' had made valuable contacts and laid the foundation for future Soviet science and technology espionage in the USA. Semyonov trained Cohen (code-named 'Luis' and later 'Volunteer') to act as his assistant, using him as a link to technical-intelligence spies he had recruited, including one code-named 'Emulsion' who still remains unidentified. By this time Semyonov's industrial espionage network included spies who were providing information to the Soviets about the manufacture of camera film and nylon, and even the miraculous new drug, penicillin.[48]

From the latter part of 1940 and early 1941 the Centre permitted Cohen to start recruiting assets of his own, and he built up a small network of agents in various military installations on the US east coast. During this time Morris Cohen's relationship with Lona Petka had deepened into a serious romance, forged through mutual physical attraction and ideological commitment. Morris's family were not overjoyed, partly because Lona was not Jewish. His brother later commented with inelegant directness that 'Moishe was a good man, but Lona always led him around by the dick.' Lona was uncertain whether to marry Morris, but a combination of his idealism and the shock of Germany's *Blitzkrieg* invasion of Russia in the summer of 1941 convinced her that they should wed. They married discreetly in a civil ceremony in Connecticut. No relative of Cohen's attended the wedding.[49]

Until this moment, although they were both members of the Communist Party, Cohen had not revealed his allegiance to the KGB to the woman who became his wife. Shortly before their marriage he had suggested her as a potential recruit to the New York KGB residency, both as an ardent communist and because she was working in an aviation armaments factory. Semyonov reported on her in favourable terms to Moscow (highlighting her evident ability as an actress), but commented that he found her a little too extrovert and emotional. The Centre approved the new recruit and, after some hesitation, soon after they were married, Cohen disclosed to Lona that he was working for the KGB. Her immediate reaction was dismay. She was eventually persuaded by her new husband to work with him for the Centre, partly convinced by the argument that in the calamitous military situation faced that summer by Russia, 'secret intelligence was [the Soviets'] principal line of defence'. The KGB assigned Lona Cohen the code-name 'Leslie'.[50]

The Cohens set to espionage work as a couple and Lona soon

proved herself a valuable asset for the KGB. She made contact, for example, with a young American engineer working in a fighter plane factory and inveigled her way into his confidence. He passed her sensitive information about the prototype of a new machine gun being manufactured in the factory, and the Centre instructed the Cohens to procure one of the weapons. According to an oft-repeated KGB story, the engineer managed to smuggle out of the factory all the parts of the gun except for the barrel, which was too bulky. To transport this without detection to the Soviet consulate in New York, which was under continuous surveillance by the FBI, required audacity and ingenuity. The engineer walked out of the factory one day clad in a heavy coat with the barrel strapped to his back to meet Cohen, who transferred the barrel to a musician's case normally used to transport a double-bass. Cohen travelled back by train to his Manhattan apartment with the case, pretending to be a musician. He then hired an unsuspecting and unemployed black man to act as an intermediary, who took the case to a flea market in Harlem where it was loaded into another car and taken to the consulate and then sent to Moscow by diplomatic bag.[51]

The theft of the machine gun proved a hollow espionage victory, because in January 1942 the US Lend-Lease programme commenced shipping enormous quantities of military equipment to Stalin, from fighter planes to tanks to guns. It was no longer a priority for the Soviets to harvest intelligence about ordinary weapons. Technological secrets became the new focus of the KGB. Over the summer of 1942, Morris Cohen recruited a French electrical engineer who was working in a research area of great interest to the Centre: radar. Cohen had no time to handle this potentially valuable agent because he had already received his draft into the US army, so he passed him to a KGB officer who had just arrived in New York, Alexander Feklisov. Over the following three years the agent, a socialist Frenchman who sympathised with Russia in its military struggle against Germany in 1941 and 1942, fed to Russian intelligence thousands of sensitive documents about radar technology. 'He was so committed,' Feklisov remarked later, 'that I often had to slow him down.'[52] Cohen recruited another source around the same time, a young radio operator and engineer called Joseph Chmilevski, who worked at a sonar laboratory on the east coast. Cohen passed the nervous and edgy Chmilevski to Semyonov, who ran him successfully until 1944 and described him in a report to the Centre in November that year as 'a highly valuable agent'.[53]

In July 1942 Morris Cohen was inducted into the American army. Classified as a 'premature anti-fascist', and destined for pedestrian work in the lower ranks, he was first to work as a cook in northern Canada and Alaska, where the USA was building the Alcan (Alaska-Canadian) highway. In spring 1944 he was shipped out to Europe assigned to a Quartermaster Battalion. After the Normandy landings his unit was engaged in transport, supply and guard duties, ending the year in Weimar and the Buchenwald concentration camp, which had recently been liberated by Allied troops. His war culminated in guard duty in Berlin before he was demobbed in November 1945.[54]

Back in New York, after her husband had been called up, Lona Cohen worked on the assembly line at two defence factories in New York. Simultaneously she acted as a courier for Semyon Semyonov's contact with the network of seven or so agents which Morris had run before he had been drafted, harvesting technical intelligence. Her task was the clandestine collection of blueprints and weapon components smuggled out of military plants in the north-east of the USA by Semyonov's agents. Another job was to make contact on the New York waterfront with US communists working on merchant ships who were carrying intelligence to New York from Soviet agents in Cuba, Mexico and South America, and to receive coded letters from an agent in America's wartime forerunner of the CIA, the Office of Strategic Services.[55] From sometime before the end of 1944 she also acted as a courier for another American Soviet agent identified as William Weisband. Lona Cohen was in contact with him through one of Weisband's two brothers, to whom Weisband mailed sensitive material in an ordinary envelope (sometimes from overseas) to be picked up later by Lona. From a family of Russian Jews from Odessa, Weisband had been drafted into the US army Signal Security Agency, the predecessor of the National Security Agency. Speaking four European languages in addition to French and his native English and Russian, he was sent to Italy as liaison officer with the French Expeditionary Corps, monitoring their messages for security violations but also stealing their military codes on the instructions of the US army (which he also shared with Soviet intelligence). Weisband was assessed by the New York KGB residency in November 1944 to have carried out 'nice work in the West [California]' and displayed 'indisputable growth during his time in the army [Africa, Italy, Britain, France]'.[56]

By this time, the Centre was already planning ahead on how best to deploy its American intelligence assets once the war was over. Semyonov was withdrawn back to Moscow. Part of his valedictory report to the chief of Soviet foreign intelligence in Moscow sketched out possible options for Morris and Lona Cohen. Cohen ('Volunteer') – still in the US army and serving in Europe – was

> fully aware of whom he's working with, he is sincerely devoted to us, ready to carry out any assignment for us. Exceptionally honest, mature, politically well versed. Ready to dedicate his whole life to our work. Upon returning from the army he should be used as our full-time illegal. He can be used along the following lines: a) as a courier; b) to select illegal operatives from among former veterans; c) to arrange safe houses and covers. He knows the restaurant business well; he could open a small snack bar that would serve as a meeting place to pass materials, letters, etc. 'Volunteer' should be given full trust.

As for his wife, 'Leslie': 'Devoted to us. No special independent work should be assigned to her for now. She could work as a courier and take care of a safe house. Later she should work as "Volunteer's" assistant.'[57] As far as Lona Cohen was concerned, events conspired to ensure that she played a more prominent espionage role than Semyonov envisaged in the closing months of the Second World War.

II

Although first alerted to the prospect of an atomic bomb in 1940, it was only by assembling clues garnered over the following years from various countries that Soviet intelligence learnt that all the resources of the Allied forces to build this weapon were concentrated in a top-secret base at Los Alamos in New Mexico. It was called the Manhattan Project because it was managed by the Manhattan, New York, district of the US army Corps of Engineers. The KGB in turn christened its priority task of penetrating Los Alamos and related facilities to obtain the secrets of the uranium bomb 'Operation Enormoz'. By the end of 1944 they had managed to develop a network of ideologically committed spies within Los Alamos and a clutch of couriers to transport their intelligence out of the clandestine research base. In October 1944, a brilliant nineteen-year-old physicist from Harvard, Ted Hall, who had recently been drafted to work at Los Alamos, offered his services to the KGB in New York while on leave. Hall (code-named 'Mlad') was concerned about the balance of power if America alone possessed the atomic bomb, and to prove his sincerity he passed over a report he had prepared on Los Alamos and the scientists working there. At the end of the year the Russians also recruited a machinist who had been assigned to work at Los Alamos, David Greenglass, and in February 1945 reconnected with an existing agent who had been sent there from Britain, Klaus Fuchs. In 2019 it was revealed that the KGB also had a valuable fourth spy in the Manhattan Project, Oscar Seborer. By now Moscow knew it was possible to build a uranium bomb and that work was progressing in America at breakneck speed to create this fearsome weapon, but lacked concrete information on how to construct it.[58]

KGB files confirm that around the end of 1944, its New York residency had renewed contact with Lona Cohen with a view to her carrying out important tasks related to 'Operation Enormoz'. 'It should be noted that she does not have experience in our work ['Enormoz'],' the Centre instructed New York, and 'You should have a series of instructional talks with "Leslie" regarding caution and secrecy in our work and also teach her a number of practical methods for checking oneself when going to a meeting, leaving a meeting.' As the significance of technical and scientific intelligence in the USA, and especially the Allies' work to develop an atomic bomb, became its top intelligence priority, the KGB had changed its personnel in the New York residency. From early 1943, the officer in charge of this line of work was the deputy station chief, Leonid Kvasnikov, assisted by Alexander Feklisov and the man who was to become Lona Cohen's most important control officer at this time, Anatoli Yatskov. Yatskov had arrived in New York in 1941 aged twenty-nine under the cover of being a new clerk at the Soviet consulate. Just under six feet tall, and with a boyish and innocent look and a shock of dense brown hair which persisted in falling across his forehead, Yatskov sharpened his image, English and tradecraft considerably after his arrival. By the start of 1945 he had developed into a well-dressed and suave KGB officer, working the diplomatic circuit with accomplished charm. He controlled some of the most important agents Moscow possessed connected with the work at Los Alamos, including the couriers Julius and Ethel Rosenberg and Harry Gold, and sources Klaus Fuchs and Ted Hall. In early 1945 Lona Cohen was used first as a safe mailing address for Klaus Fuchs to send letters rescheduling his clandestine meetings to smuggle secrets out of Los Alamos. She was also sent on her first courier runs for 'Operation Enormoz' to Canada to pick up or deliver documents in Vancouver, and even to collect a sample of uranium ore from Niagara Falls, which had been supplied by a Soviet spy in the Anglo-Canadian atomic research centre at Chalk River, near Ottawa.[59]

In May 1945, when Ted Hall used a young communist friend and former Harvard room-mate as a courier from Los Alamos, the friend was interrogated by suspicious immigration officials and the KGB were concerned that he was under surveillance. Hall suggested that Soviet intelligence should send a female courier to him in future, because it would appear more natural. The KGB agreed and decided Lona Cohen should fulfil this role.[60]

Just before dawn at 5.30 a.m. on Monday, 16 July 1945, in desert scrub near Alamogordo, 210 miles south of Los Alamos, the first test of the atomic bomb took place. The desert darkness was seared away by what Ted Hall (ten miles away in a covered rescue truck to evacuate local farms and Indian villages if the wind changed) later described as 'the cloud and the glowing, and the thing coming up and making this tremendous light'. Although aware of the impending test, Soviet intelligence urgently sought details of how the bomb was constructed to pass on to the team working to build the USSR's own weapon. Lona Cohen was hastily despatched from New York on a voyage to New Mexico. Her mission was to have a clandestine rendezvous with 'Mlad' and collect from him whatever intelligence he managed to smuggle out of Los Alamos. Her trip was destined to become part of KGB folklore. Like many tales, it was embroidered over the years in the retelling and exists in many versions, but the lineaments of the story have remained fairly constant.[61]

Lona Cohen took a train out west to Las Vegas in late July 1945 and rented a room in a resort not too distant from Albuquerque, claiming that she needed a cure for her health in the hot, dry air of the south-west. She was to meet a young man of whom Yatskov had shown her a photograph (she, of course, was ignorant of his name or identity) at 1 p.m. on Saturday, 29 July, at Albuquerque. Ted Hall did not appear. She repeated the exercise a week later with the same result. Alarmed, she sent a telegram to her home address in New York (Yatskov had organised for her mailbox to be checked every day in case of emergency), saying the contact had not come. Yatskov instructed her to wait, citing the famous saying of the French King Henry IV that 'Paris is worth a Mass'. After the young man failed to appear on the third Saturday, Cohen was increasingly alarmed, her concern heightened because her sick leave had expired. She decided to stay on for one more week and make a final possible rendezvous on the fourth Saturday after her arrival. The date, according to a contemporaneous KGB report, was 18 August 1945, twelve days after the first atomic bomb was dropped on Hiroshima and nine days after Nagasaki.[62]

Lona Cohen saw an odd-looking young man arrive wearing a straw hat, sports shirt and sandals, but most importantly carrying a yellow bag from which a magazine protruded. After some hesitation (both courier and source initially forgot the recognition 'parole'), with much relief Cohen made contact with her target. They meandered about arm in arm chatting for a few minutes. Hall handed Cohen a small sheaf of five or six sheets of paper covered in miniature scrawl. According to one Russian source, when Lona Cohen expressed frustration about him missing earlier meetings, Hall replied nonchalantly that he had mistaken the month, and that for what he had brought her that day she would have come back to meet him every day of the year.

Cohen hurried off to the train station but was aghast to find military police checking all passengers and their luggage. She retreated to the ladies' toilets, hid her railway ticket in a book and concealed the sensitive material Hall had given her under the remaining tissues in a box of Kleenex. She waited until the train was about to depart before approaching it. Two policemen started to check her suitcase and to distract them Cohen started to search in a manufactured panic for her train ticket, asking one of the men to hold the box of Kleenex. With the train whistling impatiently to leave, the ticket was finally discovered in the book, and the luggage hurriedly repacked. One of the policemen helped Cohen onto the train and handed up to her the Kleenex box just as the train was about to draw away. A less dramatic version of the incident, noted from documents in the KGB archives by Mitrokhin, has Lona Cohen encountering military police searching passengers' luggage on the train and hiding the Hall documents in a newspaper, which she asked a policeman to hold when she opened her purse and suitcase for inspection.

Back in New York, Cohen left the agreed mark on a wall to indicate to Yatskov that she was ready for a meeting the next day. Burning with impatience, Yatskov was granted special dispensation to visit Cohen at her apartment that same night to collect the information. Exhilarated but terrified, Cohen told Yatskov: 'I've never been so close to the electric chair as I was in Albuquerque five days ago.'[63] Hall's sensitive material was swiftly transferred to a Soviet intelligence officer based at the embassy in Washington and sent back to Moscow. Whatever the exact truth of the meeting with Hall in Albuquerque, Cohen displayed great presence of mind in hiding his information, and bringing these highly classified documents to New York. She was not to learn for many years their immense significance. They contained – albeit in sketchy form – plans of the first atomic bomb, including a rough diagram of its concentric spheres. It seems that Klaus Fuchs provided similar details independently at around the same time, and the Centre were able to cross-check them to confirm their reliability. The information which Hall provided, and Lona Cohen smuggled to the Soviets in August 1945 at risk to her life (if caught she could have faced the death penalty for espionage in wartime), was a key foundation for the USSR atomic bomb project which was to sprint forward in the coming years.

The defections of Gouzenko and Bentley in late 1945 alarmed the KGB and caused the Centre to instruct that all contacts with sensitive sources should cease. During the late 1950s, the FBI discovered that the Cohens had visited France in 1947 but did not discover why. KGB files provide the explanation. In 1947 a message had been sent to the Cohens instructing them to travel to Paris to meet their two former controllers, Semyonov and Yatskov. They and their 'Volunteer' network of spies were to be activated again and handled by a new KGB officer in New York, Yuri Sokolov. Lona Cohen – who had stayed in touch with Ted Hall – passed on some intelligence about

Hall which was unwelcome to the Russians: he had become publicly involved in 'a peace movement and wanted the bomb to be banned'. Russian intelligence much preferred 'Mlad' to remain a valuable but always low-profile source.[64]

Sokolov made contact with the Cohens in New York in the autumn of 1947. The Cohens first re-established links with some of their 'Volunteer' agents, who provided new information about radar, so-nar and radio guidance for missiles, and, according to some Russian sources, included the same agent who had worked during the war in the Office of Strategic Services, and had since been recruited into the fledgling CIA. Meanwhile, alarm had been triggered in Moscow by intelligence reports of the Americans developing a hydrogen bomb with many times the destructive power of the A-bombs dropped in Japan. The research for this project was centred at the University of Chicago, which in turn became a principal intelligence target. By this time Ted Hall was in his second year of postgraduate study in physics at the same university, but had decided that he no longer wished to provide information himself for Soviet intelligence. When Morris Cohen met him in summer 1948 he failed to persuade Hall to change his mind, but Cohen reported to the KGB that he had ob-tained 'important information on "Mlad"'s two new contacts. They have declared their wish to transmit data on "Enormoz", subject to two conditions: "Mlad" must be their only contact and their names must not be known to officers of "Artemis" [Soviet intelligence].' In the late 1940s Sokolov used to meet Lona Cohen about twice a month and Morris less often.

Much to the chagrin of the Centre, Hall raised his personal pro-file even further by joining the US Communist Party that autumn. The Centre partly blamed the Cohens for this development because they had not followed instructions, failing to maintain 'regular contact' with 'Mlad' and to set up a cover for themselves (the plan was for them to open a photographic studio). Nonetheless, KGB archives confirm that the 'Volunteer' network widened to encom-pass at least two new agents, code-named 'Aden' and 'Silver', who were undoubtedly the two nuclear physicists contacted by Hall. Their identities remain unknown, but according to some sources they were a married couple, and at least one of them worked at the plutonium-making complex at Hanford, Washington. It seems from Soviet sources that sometime between 1947 and 1950 agents run by the Cohens provided valuable information to Sokolov to help USSR scientists develop a hydrogen bomb – intelligence which the Centre rated highly.[65]

Around this time, because of the chilling onset of the Cold War and suffocating surveillance by the FBI, the Russians decided to deploy Vilyam ('Willie') Genrikhovich Fisher – later better known under his pseudonym Rudolf Abel – as an illegal in New York to rebuild intelligence operations in America. Fisher arrived in the USA in November 1948, and early the next year was given the Cohens and their 'Volunteer' network as the foundation of his illegal residency. This group, according to the official history of Russian intelligence, transmitted to the Centre 'super-secret information concerning the development of the American atomic bomb'. Some of this sensitive material probably concerned a top-secret American industrial innovation to allow the mass production of polonium 210, a key ingredient in the trigger mechanism of a nuclear bomb. In acknowledgement of the value of the intelligence collected by the Cohens' network, in August 1949 Fisher was awarded the Order of the Red Banner. This was the same month that the USSR – on the 29th – successfully exploded its first atomic bomb, which was almost identical to the one dropped on Hiroshima and Nagasaki. Three days later a specially equipped United States air force weather reconnaissance aircraft flying from Japan to Alaska collected atmospheric radioactive debris of the Soviet test. This news shook the Western powers: American intelligence considered that the Soviets would not produce an atomic weapon until 1953, while the British did not expect it until 1954. There was now theoretically a balance of atomic power and this further strengthened Ted Hall's resolve not to provide sensitive information about the American atomic programme to the Russians based on his own work. The Cohens (and, in all likelihood, Willie Fisher too) were ordered to convince Hall to change his mind, but he would not relent. In recompense for past services, however, the Centre instructed Lona Cohen to deliver a gift of 5,000 dollars to Hall in Chicago.[66]

There is evidence that Lona Cohen travelled to the US–Canadian border after the war on behalf of the KGB. One of her contacts there was almost certainly Bruno Pontecorvo, the Italian physicist and KGB spy, who was working on the Canadian NRX nuclear reactor at Chalk River. Pontecorvo was at the heart of the Canadian project, which was producing novel forms of uranium and other elements essential for nuclear weapons and power. Pontecorvo travelled regularly from Montreal to the US border under the cover of needing to do so to keep his application for US citizenship active. The reactor started operating from 1947, and it seems highly likely that Lona Cohen received from Pontecorvo a sample of uranium from there in late 1947 or 1948 for transmission to Moscow. It is also possible that Pontecorvo passed to Lona Cohen blueprints of the Canadian reactor. Pontecorvo defected to Russia suddenly in 1950. The important KGB defector, Oleg Gordievsky, told MI6 he knew Soviet intelligence officers who had had dealings with Pontecorvo and they told him the Italian had been a hugely valued agent during and after the Second World War. Lona Cohen also acted during this time as a courier for the American spies Julius and Ethel Rosenberg, who were ferrying other secret information to Soviet intelligence.[67]

In late spring 1950, the KGB watched with mounting concern the arrest of its network of spies which had penetrated Los Alamos. Fearing that the detention of the Cohens in the USA might lead to yet more damage, the Centre ordered their exfiltration on 16 May 1950. Such was the urgency that Sokolov took the risk of visiting the Cohens at their apartment to pass on the order, exchanging written notes with them in case the property was bugged. The KGB prepared Mexican passports for them and in July the Cohens took a boat from New York to Veracruz in Mexico. Cohen was heartbroken to leave his father. 'Emotionally I couldn't bear it and I almost missed the ship,' he recalled later. 'My father understood that we would never see each other again.' In Mexico they were hidden for several months by two Soviet agents, both members of the Spanish Communist Party in exile. Moscow Centre despatched its best forger from Directorate S, the legendary Pavel Gromushkin (who first learnt his trade in the 1930s airbrushing faces out of photographs during Stalin's Terror), to Veracruz to prepare a new set of forged identity papers for the Cohens. It was as an American couple named Briggs that they crossed the Atlantic in a Polish steamship to Amsterdam, then travelled to Prague by train, before finally being flown to Moscow. Here, in a final irony, as a result of the sort of bureaucratic shambles which can afflict an intelligence service as much as any other branch of government, there was no one from the KGB to meet the Cohens at Vnukovo airport. The Cohens passed through the security checks with their fake US passports. Outside they were asked twice – first by a taxi driver and then the driver of an official Intourist travel coach – if they wished to be taken to the American embassy. Unsurprisingly, they declined.[68]

The Cohens were to be based in the Russian capital from 1951 to 1954, although the KGB was later to spread disinformation that they were living in Poland. Ill at ease in Moscow, the Cohens barely left their apartment for many months, visited almost exclusively by their KGB liaison officer. Bored and no doubt a little depressed by their isolated existence in the grim Russian capital of the early 1950s, the Cohens pressed regularly to be allowed to work again for the Centre. Revealingly perhaps, when asked by the KGB why she was so committed, Lona blurted out, 'Probably it's because secret service work is for us like heroin for drug addicts. For the moment we can't imagine our lives without it.' Eventually, Soviet intelligence concluded that the Cohens were too valuable a resource to waste. The first plan was to send them to South Africa (hence the origin of the cover name Kroger), but the Centre decided instead to deploy them in England. Their 'legends' as the New Zealand couple Peter and Helen Kroger were elaborated and the couple trained in espionage tradecraft and clandestine communications. In March 1954 they travelled to a small town near Vienna to send off to Paris for their two New Zealand passports in the name of Kroger with the help of Soviet agent 'Paddy' Costello. After returning to Moscow for some final briefing, on 3 December 1954 they departed the snow-covered Russian capital for the rain and smog of a London winter.[69]

Within a few months of their arrival, Morris Cohen collected their Canadian 'escape' passports from a dead letter box in a male toilet in the Odeon cinema in Leicester Square. After the brief meeting with Molody in Paris in early 1955, it seems the Cohens re-established contact with him later that year in the Lyons Corner House café near Piccadilly. According to one Russian source, the Cohens had only been told they were to meet there the new KGB illegal *rezident* in London, Gordon Lonsdale. They had no idea who Lonsdale was or what he looked like (they only knew Molody as 'Ben' or 'Arnie'). The Cohens (whose new code-name collectively was the 'Dachniki', or holidaymakers in Russian) arrived a little early to check there were no acquaintances there whose presence might compromise the clandestine meeting. Lona Cohen immediately recognised Molody and pointed him out to her husband. He urged caution and suggested they walk out into the hall and wait for a signal. As they ambled past, Molody greeted them with his usual bonhomie, hugged Lona and introduced himself as Gordon Lonsdale. After confirming their mutual identities through the exchange of 'paroles' ('Did I last see you in Mexico City?'; 'No, we could only have seen each other in Ottawa'), they discussed business and the Russian passed the Cohens a miniature camera. At the start of 1956, with Molody's approval, the Cohens bought 45 Cranley Drive. Molody started to visit the bungalow around this time, and on one occasion in their garden could not resist drawing attention to himself by giving a loud 'wolf-whistle' at the very attractive Italian wife of a neighbour. During his overnight stays in Ruislip, Molody helped Morris Cohen dig out debris in the cellar and create the hiding place for the Cohens' first transmitter. Cohen collected this Astra radio in Brussels in about October 1956, a more primitive version of the advanced model found at the bungalow after the arrests. Molody, as the illegal KGB resident in London, was to develop and control a network of agents. The Cohens' main role was to act as his communications operators but also help KGB spies in the embassy, receiving instructions from Moscow via radio, send and receive sensitive material to and from dead drops and dead letter boxes and by using microdots, and to scout for possible recruits. Some of the receptacles for secret messages engineered in Moscow were elaborately camouflaged, like a piece of rotten wood with a hollow interior collected by Morris Cohen on one occasion and prised open with a screwdriver.[70]

It seems from the KGB archives that the Cohens began their espionage work in 1956. They helped recruit a former MI5 officer code-named 'Baron' who provided useful intelligence about the Suez Crisis. They also handled material from Houghton at Portland (then controlled by a legal KGB officer in the Russian embassy), which on one occasion they were instructed to turn into microdots and send to the 'usual address'.[71]

Although Molody and the Cohens were illegals they liaised with a KGB Line N (illegal support) officer based in the London embassy, who supplied them with money and instructions from the Centre, and (in Molody's case) letters in microdot form from his family in

Moscow, delivered via dead drops and face-to-face meetings. Using KGB funds, as MI5 had discovered, Molody set himself up as director of several companies dealing in jukeboxes, vending machines and one-armed bandits. A KGB file suggests the vending machines included bubblegum dispensers at numerous sites, giving Molody an excellent pretext for journeys in the London area to meet his Line N support officer, the Cohens and his agents. At all times, of course, the three illegals walked on a knife-edge: a loose word, their identity revealed by a defector, an unlucky encounter with an old acquaintance, and they faced exposure. Molody later recounted one such heart-stopping incident. He was passing through Le Bourget airport in France when he was hailed in Russian by a friend from his days at Moscow University. Molody stared at him coldly. Nonplussed, the friend said in English, 'I thought you were someone I knew, Konon Molody.' 'No, you made a mistake,' the illegal responded crisply in English. The friend was convinced the man before him was Molody and was about to ask again when Molody leant forward and whispered in perfect colloquial Russian, 'Piss off to your ****ing mother, you cretin.'[72]

Speculation has swirled since the Portland trial about whether the network of spies handled by Molody and the Cohens was wider than Houghton and Gee. Radio transmissions from Moscow to the UK were intercepted by GCHQ for eleven days after the arrests using call-signs found in Molody's and the Cohens' signal plans, and were assumed to be sent to other KGB agents in Britain, but there was no other evidence of additional spies found at the time. When interviewed in prison by Charles Elwell, Molody referred to another agent he controlled but he never provided any details which the Security Service could investigate. Even former MI5 officers familiar with the case in the early twenty-first century express scepticism that the KGB would have invested so much time and effort in forging an illegal espionage network to control just two agents, Houghton and Gee. Evidence from Russian and other sources demonstrates that the size and importance of Molody's espionage ring were indeed greater than previously thought. It has been known for some time that, in addition to Houghton and Gee, Molody controlled Melita Norwood (the so-called 'Granny Spy' exposed as a long-term Soviet agent in 1999) for around two months at the end of 1958 before she was passed back to the legal residency – there is speculation that the two did not gel because the ideologically puritan Norwood scorned Molody's playboy image. There were certainly other agents in Lonsdale's network.[73] This was confirmed by former KGB officer Vasili Dozhdalev, who controlled Molody in London and was later a senior figure in the KGB illegals department: 'MI5 did miss some people [when Molody was arrested]. But if it worked longer on the case it would have missed Lonsdale as well. It is hard to tell what would have happened if MI5 had let the case run on longer. Probably they would have had more information, more details. But there were some people who escaped [arrest] even as it was, so probably even more people would have escaped.'[74]

One key target, for example, which it appears the KGB successfully penetrated was Britain's chemical and biological weapons research centre at Porton Down. When the Second World War ended, the Allies were amazed by the advanced state of German technology regarding nerve agents such as tabun, sarin and soman, which the Nazis had developed but fortunately never used. Nerve agents are highly potent, odourless and colourless, and work rapidly through the skin, eyes and respiratory tract. If a state possessed chemical and biological weapons it was provided with an additional and uncertain military advantage; and it was crucial to know how to defend against them. Further, it was known that the USSR had rebuilt one of the German chemical weapon plants at Beketovka, near Stalingrad, after the war and undoubtedly possessed a nerve agent capability.

With the onset of the Cold War, the UK, USA and Canada researched new nerve agents, and scientists who had worked in Germany during the war were interrogated and recruited to assist. In 1952 two ICI scientists in Britain synthesised a new insecticide with devastating side-effects on humans. Porton Down scientists secretly researched this lethal substance and, based on their work, the so called 'V-series' of nerve agents were developed (V standing for 'venomous'), including the highly toxic VX – a thick, oily and odourless liquid with a terrifying ability to kill. A drop of VX on the skin could kill a man in fifteen minutes, and a litre of the fluid theoretically contained sufficient individual lethal doses to kill one million people. In February 1957, the US army Research and Development Command chose VX as the agent on which to focus further work. Development of nerve agents at Porton Down was cut back after the Cabinet Defence Committee took the top-secret decision in July 1956 to have a solely defensive chemical weapons capability. In reality, as the scientists involved understood well, the chemistry involved was relatively simple, and if a state possessed an effective defensive capability in this area, it could quickly manufacture an offensive one. Research work therefore continued, in cooperation with the USA and Canada, on nerve agents and on pathogens and bacteria used in biological warfare, centred on the Microbiological Research Department at Porton in the late 1950s and 1960s. During this period Porton also developed CS gas, used to control riots.[75]

The Centre instructed Molody as a priority to penetrate Porton Down to ensure that the USSR could compete with the West in the sphere of biological and chemical weapons. According to one Russian source who had access to KGB files, the message to Molody from the head of the KGB's First Chief Directorate ordered him to focus in particular on 'a certain lethal substance which Porton has developed, of which 200 grams would be sufficient to kill the population of the globe'. Russia's abiding interest in chemical and germ warfare had begun around 1920. Lenin commanded the establishment of a secret 'Laboratory X', involved in developing and testing undetectable poisons and toxic products, used to kill dissidents, political opponents, double agents and defectors. This was transferred to the KGB in 1937. At the time Molody and the Cohens were working in England, the KGB's chemical and biological espionage work was focused in Directorate S because the intelligence-gathering and other related work (such as sabotage) was often carried out by illegals. Russian sources indicate that Molody and the Cohens compiled dossiers on the background of staff at Porton with a view to recruiting an asset, and Molody was ultimately successful, obtaining samples of biological weapons, including lethal bacteria, and probably of chemical weapons like VX, and detailed intelligence about CS gas and exchanges with the Americans. A sample of one (unidentified) highly lethal bacteriological weapon, it seems, was sealed in a test tube in a thermos flask specially adapted for the purpose by the Centre. This was passed to the Cohens, who concealed it in a dead drop at Highgate in north London, to be collected later by a legal KGB officer from the embassy for despatch to Moscow. The Cohens used their radio to send a coded message from

Molody to the Centre in August 1958 to alert them to this espionage success.[76]

Alexander Kouzminov, who worked in this area for the KGB until his defection in 1992, has confirmed that until their arrest in 1961 'these illegals [Molody and the Cohens] gathered and transferred to the Soviet Union Britain's most valuable secrets about the creation of biological weapons', and that the KGB had a long-term penetration agent operating in Porton Down under the code-name 'Rosa' (whose identity and date of recruitment are unknown). A secret 1960 British intelligence report on Soviet biological warfare concluded that the British 'must assume Soviet parity with the West in basic knowledge and technical know-how, and familiarity with the broad lines of current Western outlook on B.W.', while one on chemical warfare stated that '[it] is certain that Soviet research . . . is of such standard that it is likely by now to have resulted in the development of agents similar to our V-agents. Production of V-agents could by now be in progress.' The authors of these reports could have had no inkling that Soviet research and development in these areas may well have been boosted by KGB penetration of Porton Down. Throughout the 1960s the USSR continued to develop its chemical and biological weapons, and in 1969 a report to Britain's Joint Intelligence Committee confirmed that the USSR possessed 'a highly toxic filling known as VR-55 which is at least as effective as the most powerful nerve agents known to the West'.[77]

The official history of the KGB discloses the existence of another agent, 'K', who was controlled by Molody from 1958. Details are sketchy. It appears 'K' had access to valuable intelligence about new weapons production and related technologies, and had various 'brush contact' meetings in London with Molody – sometimes in town and sometimes on a train – to pass over documents, which the illegal resident urgently copied and returned. This information included sensitive details about the materials being used in the construction of the nuclear submarine, *Dreadnought*. The official history asserts that the Centre valued the secrets divulged by 'K' because – as with the intelligence stolen by Houghton and Gee – they 'helped the USSR plan rational military programmes, saving much money'. Although it is not clear what his sources were, Molody also garnered new scientific and technological intelligence about important industrial operations: for example, rolling high-strength aluminium (vital in aircraft construction) and how to join titanium to stainless steel (important in the chemical industry and in building sections of nuclear reactors and aircraft engines). It appears he obtained sensitive information about Britain's programme to build an independent medium-range ballistic missile, Blue Streak. The design was complete by 1957 but the project was cancelled in 1960 because of rising costs in favour of the American Skybolt. One Russian source suggests this intelligence was 'highly valued by Soviet scientists'. Molody's posthumous memoirs in Russian indicate that military and civil uses of nuclear energy were another priority area where he enjoyed some success, weaselling out intelligence, for example, on the secret exchanges between the UK and USA about nuclear weapons and nuclear-powered engines for submarines and surface ships, and copies of highly sensitive lists of intelligence targets in the USSR elaborated jointly by MI6 and the CIA. Molody's sources for this intelligence remain unknown. His (unreliable) 1965 memoirs refer to him having two illegal assistants in Britain: one code-named 'Wilson', who helped him collect intelligence, and someone used for radio communications with the Centre on regular 'operational matters', who operated a transmitter separate to the one at 45 Cranley Drive (which Molody suggests was used 'solely for emergency'). In conversation with Charles Elwell in prison, Molody also mentioned 'an auxiliary illegal network' he worked with. It is possible that Molody did have such assistants. There are numerous references to 'Wilson' and his supposed central role in penetrating Porton Down in a book about Molody published in Moscow in 1990, partly based on interviews with Molody just before his death, which refers to one source of information for the KGB within Porton being Dr Geoffrey Bacon, a scientist who was accidentally infected there with and killed by the plague virus, *Yersinia pestis*, in August 1962.[78] There is no evidence in other sources to support these claims, and no evidence at all that Dr Bacon was ever an agent of Russian intelligence.

Molody's assessment of the work of the Cohens for the Centre dated 13 August 1958 was full of praise:

The 'Dachniki' are of considerable help to the illegal *rezidentura* in that they take responsibility for various espionage operations relating to the collection of intelligence coming from Porton and Portland. The 'Dachniki' [referring to Morris Cohen] is also assisting the *rezident* in selecting and dealing with potential recruits. He has established fruitful contact with 'Chambers', an employee of the services of the British army; and also with 'Elliott' and 'Baron', two former officers of MI5.

The Centre was duly impressed by the Cohens' performance and four days later the head of the FCD asked for the necessary paperwork to be drawn up to award the Cohens the Order of the Red Banner. Remarkably, Molody had not as yet been passed control of Houghton and Gee. It was this fateful move, taken in Moscow only in the summer of 1959, which was to lead to the downfall of Molody and the Cohens.[79] Before discovering from the KGB archives why that decision was taken, it makes sense to trace what happened to Houghton and Gee after they were freed from prison.

15

Houghton and Gee – life after prison

When they were released from prison in May 1970, Harry Houghton and Ethel Gee may have wished to be left in peace, but it was not to be. There could not have been a sharper contrast between the lives of Molody and the Cohens after their return to Moscow, sheltered from any media intrusion, and the rumbustious way the British spies were pursued by journalists of a free press. The news cameras caught the outwardly calm Gee returning to her former terraced home in Portland to be greeted by a milling crowd of flash bulbs, newspaper men and locals, some craning from upstairs windows to glimpse the homecoming of their former neighbour. She told waiting reporters initially that she did not want to see Houghton again and would not welcome it if anyone brought him to her home.[80]

When she walked the surrounding streets in the days that followed, with her neatly permed hair, wing spectacles and pale raincoat, she was greeted with hostile stares and the occasional cry of 'Traitor'. Houghton stayed in a hotel in Bournemouth and sent a message to Gee requesting a meeting, accompanied by a bottle of sherry. Gee agreed and was smuggled out through the back gate of her house and driven to a hotel in the New Forest, where they posed for photographs and were interviewed together. With hollow eyes and an air of resignation, Gee commented, 'I just want to lead an ordinary life,' with Houghton – his receding hair slicked back and face lined – cutting in, 'If they'll let you . . . It'll be a nine days' wonder.' After their first encounter after release from prison Gee remarked that it was 'like meeting a stranger'. In a separate interview, Houghton confessed that he had spied 'for money', but refused to disclose how much he had been paid. Houghton started to visit Gee clandestinely at her house after dark, entering by the passage at the back of the house. The couple clearly found comfort in each other's company, talking in confidence about the truth of their espionage, because in April the following year they married secretly in a civil ceremony. Houghton was then aged sixty-five, Gee nine years younger, and they lived together in a house they bought in Poole.[81]

As Houghton had forewarned MI5 when interviewed in prison, after his release he would be keen to make cash from his memoirs. He published this tiresome volume in 1972, *Operation Portland*, a book-length whine of complaint about the alleged incompetence of the Security Service, the unfairness of his trial and his harsh treatment in prison. He showed little or no insight or self-knowledge, suggesting that 'a deep bond of friendship existed' between him and Molody (who had made clear his scorn for Houghton when talking to Elwell in prison), and displayed an unbridled willingness to invent fanciful stories for sensation (for example, that he helped the KGB land secret agents from submarines in the dead of night on the coast near Portland wearing 'Wellingtons and . . . their shoes secured round their necks'). As a contemporary review in the *Spectator* mused, 'One of the difficulties about this book is to know how much of it to believe.'[82]

After publication of Houghton's memoirs, the couple dropped out of public sight. They ran their home as a guesthouse for a period but then retired to a three-bedroom house on a nondescript 1970s housing estate at Fleetsbridge, near Poole in Dorset. In 1984 Ethel died aged seventy, and Harry followed her the next year on 23 May, just before his eightieth birthday. In his will, Houghton stipulated that his pug dog Ross 'be destroyed as soon after my death as possible', and left various minor bequests to friends and neighbours; but with no descendants (he was estranged from his daughter from his first, unhappy marriage decades before), and, as if to make a gesture to atone for past mistakes, he bequeathed the vast bulk of his estate (worth well over £300,000 in today's money) to a local hospital cancer charity. During their lives, Houghton and Gee never revealed the truth of their espionage. It was over a decade after their deaths before the truth – or at least as much of the truth as the KGB would sanction for release from their archives – emerged.

16

Houghton and Gee as spies

In about 1996, a retired KGB colonel, Oleg Tsarev, was allowed access to the KGB archives and permitted to bring out each day a certain number of pre-vetted files about Soviet spies in Britain to the press bureau of the SVR near the Lubyanka. The files included those on 'Shah', the code-name for Houghton. Here Tsarev and the English spy writer known as Nigel West were allocated an office and an L-shaped desk. Since West knew no Russian, Tsarev summarised the content of each document, and because of the limited time available, West was forced to decide on the spot whether it should be copied for him or not. The photocopier in the press bureau was, in a predictable reflection of the collapse of many aspects of Russian society at that time, broken. Tsarev was compelled to smuggle the 'Shah' documents to be copied out of the building each night to the Marriott hotel a few blocks away, where capitalism could be trusted to provide a photocopier that worked, before infiltrating them back into the precincts of the SVR press bureau ready for the next day's work. These hitherto top-secret documents provided for the first time a reasonably accurate account of Houghton and Gee's spying, and nailed many of the lies they had told at their trial and which Houghton had repeated and embellished in his memoirs.[83]

As the jury at the trial had rightly suspected, Houghton had not been coerced into espionage. It was he who took the initiative. In an ironic echo of how Goleniewski had first approached the CIA, soon after his arrival in Warsaw towards the end of 1951 Houghton sent a letter to the private secretary of the Polish Minister of Foreign Affairs, his white envelope closed with five black wax seals. The missive said it was thought the secretary to the British naval attaché in Warsaw was willing to supply secret information in exchange for money. Although unsigned, the letter gave the address of a flat in Warsaw. When a Polish intelligence officer visited the flat on

19 January 1952 he found a man, Houghton, who did not admit to writing the letter but agreed to be an intermediary and pass on information from the embassy in exchange for money. Houghton told the officer he thought 'Britain's present rulers had sold the country to the Americans and turned it into an American colony'. He entered the employment of Polish intelligence under the code-name 'Miron'. The secretary to the naval attaché soon streamlined the supply of documents so efficiently that the Soviets were frequently reading sensitive naval papers before their intended embassy recipients, because the diplomatic mail arrived by special courier plane each Wednesday from Berlin and Houghton provided documents to his Polish contacts that same night to be photographed. He supplied 715 documents in May 1952 alone, rising to 1,167 in August. The sensitive material he stole in those early months included an official estimate of the Polish armed forces, the complete structure of Naval Intelligence in London, a naval summary alerting the KGB to a possible British spy operating at a Soviet naval facility in Murmansk, and ten codebooks used by British military attachés. The Centre assessed this material as 'valuable' and 'very valuable'.[84]

When it seemed likely that Houghton would be transferred back to London early, control was passed to the Centre. Moscow, wary of alarming such a valuable source, decided to pretend that the operation was continuing to be run by the Polish intelligence service. In September an experienced KGB officer masquerading as a Pole, Alexander Feklisov,[85] was sent from Moscow to Warsaw to meet Houghton and assure him that contact would be established in London. A cable was sent to the London residency alerting it to Houghton's impending arrival. In February 1953, Houghton met a London KGB officer outside the Dulwich Picture Gallery and, after exchanging the agreed 'paroles', revealed that he had started work as an administrator at UDE in Portland but was doubtful he would have access to sensitive material at all comparable to that he exploited in Warsaw. The KGB man advised Houghton not to be despondent about the opportunities and advanced him £50. His optimism was not misplaced.

At their next clandestine meeting Houghton revealed he had worked out a method to gain access to the 'safe room' where classified documents were stored: it seems he had won the confidence of the woman in charge of the depository, who asked him to take over when she went to lunch, allowing him to range freely over the material held there. Suspicious of such a blatant breach of security, the Centre harboured doubts about Houghton's reliability for some time, but they were finally convinced by the authenticity and industrial quantity of the documents he furnished: 1,927 pages in 1954 and 1,768 in 1955. They included secret Royal Navy tests on new devices to conceal propeller noise, deflect torpedoes and detect high-speed submarines, and sensitive Admiralty orders to the Royal Navy (sufficiently important that they were circulated by the Russians to its Soviet equivalent). The Centre was further impressed early in 1955 when a new KGB officer took over as Houghton's handler and the Englishman smuggled out a subject catalogue of secret documents. As a result, Moscow could in effect order individual documents at will and the Admiralty clerk was rewarded with substantial cash bonuses (£500 in December 1955, for example).

All proceeded smoothly and lucratively for Houghton until security concerns were raised about him in late summer 1956 (culminating in his transfer to the naval dockyard – as MI5 had discovered during their 1960 investigation). Houghton, of course, did not wish to undermine his credibility with his Soviet handlers by revealing the genuine reasons for his transfer, including the rumours circulated by his wife, and misleadingly presented the move to the dockyard to his KGB controller as a promotion. The main consequence, however, which seems to be borne out by Houghton's KGB file, is that his access to sensitive material dried up to a great extent. The following year, in October 1957, after extensive consultation with the London *rezidentura*, the Centre determined that the English spy should finally be told that his case had been transferred from the Polish intelligence service to the KGB several years previously. Houghton had guessed the truth long before, commenting, 'We should be able to work together even better.'[86]

'Shah' mentioned Ethel Gee to the KGB for the first time in early 1955. With his marriage collapsing, she had become his girlfriend, and she accompanied him to London when he travelled there to meet his KGB handler. Alarmed that any approach to Gee for assistance by Houghton might precipitate his exposure and the loss of a valuable asset, the Centre warned Houghton not even to hint at his covert activities. The Englishman seems to have complied with the Centre's prohibition and only raised the issue again in October 1957, when he said he was considering marriage to Gee, and ideally would like to recruit her. Moscow was suspicious, apprehensive that Houghton had already brought Gee into his confidence and was seeking retrospective approval from the Centre. The wheels of decision-making turned ponderously in the Lubyanka and it was not until a year later that the Centre finally consented to Houghton approaching his girlfriend. According to Houghton, Gee's first response in January 1959 was delight that he was not meeting other women, closely followed by disbelief and tears. When calmer, she agreed to keep Houghton's secret but refused 'to take part – she is terribly afraid of the possible consequences'. Gee appeared obdurate, but when Houghton next met his KGB handler in London two months later he produced a bundle of documents about the anti-submarine Super ASDIC 184 sonar, which had been passed to him by Gee. Houghton must have been as overjoyed as the Centre: after a relative drought of access to sensitive material for the previous two years, the stream would once again start to flow. Gee was code-named 'Asya'. A month later Houghton was allocated a new KGB controller from the Russian embassy, the able Vasili Dozhdalev, who had recruited George Blake in Korea in 1951.[87]

Ironically, the KGB files on 'Shah' reveal that it was the accomplishments of the KGB in Britain which triggered the chain of events that led eventually to the crushing of its successful illegal network headed by Molody. On 25 May 1959, the head of the FCD, Alexander Sakharovsky, wrote to the service's chairman expressing concern that the London residency was the object of intense surveillance. He recommended that control of some agents should be transferred from officers in the embassy to Molody's illegal network, which was not limited by the same travel restrictions as diplomatic staff: 'Taking into consideration the existence in Britain of a number of very valuable agents, contact with whom from the position of the London *rezidentura* is unsafe, I should consider it expedient to pass over some of the agents to "Ben" [i.e. Molody]'s *rezidentura* . . . in the first place "Shah" [Houghton] . . . who gives valuable documentary information.'

The KGB Chairman approved the recommendation by marking it with two letters, 'ZA', meaning 'I am for it', and Molody was introduced to Houghton by Dozhdalev in July 1959. Dozhdalev was later to criticise bitterly this decision to transfer control of Houghton, but at the time its fateful consequences could not be foreseen. When Molody met Houghton the next month he suggested he should visit him at his cottage to demonstrate how to operate a 35mm camera and develop film. After some delay to work out an appropriate cover story, the overnight visit took place in early January 1960. Molody taught Houghton some basic photography and was introduced to Ethel Gee. Molody's note of the meeting underlines that Gee knew precisely what she was undertaking and for whom: '"Asya" knows well who we really are. She's very friendly and talkative, but an awful cook. "Asya" had and has access to documents of all completed projects ... in her establishment including blueprint copies. She can get a spare copy of the majority of projects which at present are being worked out. The difficulty is that (a) she doesn't know what to take out and (b) how to take them out.'

Molody advised her how to evade the UDE security controls and later gave her a list of items in which he was interested.[88] At Houghton's next meeting with Molody in June 1960 (MI5 knew he visited him at his cottage at this time), he handed over various documents listed in the UDE index of underwater detection equipment, including, significantly, ones about the new Type 2001 sonar designed for the *Dreadnought*. The next two meetings, on 9 July and 6 August, were monitored by MI5; but on 1 October, the meeting when Security Service surveillance lost contact with Houghton, he in fact met Dozhdalev and passed him a roll of film. At some time – the date is unknown – Houghton also passed to the Russians details of secret hydro-acoustic stations, underwater listening posts.[89] KGB files refer to the next two meetings, including the one of 18 December when, intriguingly, Molody asked Houghton to bring Gee along to the next rendezvous, planned for 7 January 1961, because he had still not answered some of the questions Molody had posed back in August. It was at that meeting, of course, that Special Branch seized the three spies.

If the evidence contained in the KGB files about Houghton and

Gee had been revealed at their trial, there is little doubt that the Lord Chief Justice would have imposed a more severe penalty on the pair, perhaps fifteen or twenty years rather than ten years in prison. This may help to explain Gee's obdurate refusal to cooperate in any way with MI5 while she was in jail.

Epilogue

After Lona's death in 1992, all meaning leached out of Morris Cohen's life. Increasingly frail, his memory, hearing and eyesight weaker with each passing month, he survived her by only two and a half years, dying in an intelligence service hospital in Moscow on 23 June 1995 aged eighty-five. At the close of a hot summer day he was buried in the KGB's Novokuntsevo cemetery adjacent to his wife. A guard of honour presented arms and fired a salute over the grave, causing the birds in the surrounding pine trees to screech and scatter. The following month, by order of President Boris Yeltsin, Morris Cohen was posthumously awarded the title of Hero of the Russian Federation.[1]

Morris Cohen was the last of the Portland spy ring to die, and the last of the three living KGB witnesses who knew the answer to an abiding riddle about the spy ring: did Molody and the Cohens know they were compromised before their arrest and warn Moscow? While in prison Molody told Elwell that he had become suspicious, changed his ciphers, reported his concerns to Moscow, but was instructed to stay in London. He said his suspicions were partly based on the fact that a watch had disappeared from his Regent's Park flat and there was an attempted break-in at the Cohens' bungalow just before Christmas. Elwell was clear that neither of these incidents was linked to the Security Service. This hare was set running in public in Molody's notoriously dodgy 1965 memoirs, when he said he knew his possessions in the bank safety deposit box had been searched while he had been away on holiday in the early autumn of 1960, and also his London apartment. KGB sources repeated variations of these tales. In a KGB interview in 1989 Morris Cohen said his suspicions had been aroused by spotting a green Post Office van in the vicinity of their house in December 1960 (the month GCHQ carried out their only attempt to intercept radio signals transmitted from 45 Cranley Drive). Four years later the retired KGB officer Vasili

Dozhdalev stated that Molody had been alerted that his espionage work at Portland was coming under MI5 scrutiny, but his contacts could not indicate how advanced the Security Service investigation was. Some sources in Moscow have even suggested that Molody was so concerned that he asked permission from the Centre to be exfiltrated, but the KGB delayed too long and did not give its approval before the arrests.[2]

On the other hand, the SVR official history and other accounts by former KGB officers such as Vitali Pavlov make no mention of any suspicions on the part of Molody or the Cohens before 7 January 1961, and when Molody was back in Moscow lecturing to young recruits he told one of them that his arrest was unexpected. That same KGB officer, Mikhail Lyubimov, in retirement after years of experience abroad, has stressed that no Russian illegal would ever risk meeting an agent if he thought he was under surveillance. The risk of compromising that source was too heavy. KGB tradecraft required the illegal to inform the Centre and escape.[3] The truth was probably as messy as the conflicting claims suggest. Like all KGB illegals, Molody and the Cohens lived under the permanent shadow of being exposed and were watchful and suspicious. It is quite possible that Molody's doubts were aroused in the last few months or weeks of 1960, perhaps by some clumsy surveillance or even a vague tip-off from a Russian intelligence source. Molody would not have organised a meeting with Houghton if he genuinely believed that he was under surveillance at the time. It is probable too that, since the arrests were a severe embarrassment to Moscow, an element of their claims was disinformation: a desire by the Centre to deflect criticism from itself and onto the British intelligence services. It is, of course, impossible to winnow out what really happened without access to the KGB archives.

Another riddle about the Portland spy ring can now be answered with more certainty: how much damage did it cause to the West? The KGB files on Houghton describe comprehensively the nature and quantity of the documents he (and later with Gee) disclosed to the Russians: in total around 17,000 pages, according to the SVR official history. Up to 1957 these included naval attachés' codebooks, secret Admiralty orders, and papers on various sensitive detection and counter-detection devices; and from March 1959, at the very least, intelligence about the anti-submarine Type 184 (used by surface ships to search for submarines and torpedoes) and documents about various underwater detection devices, including in particular the Type 2001 sonar.[4]

This revolutionary sonar was specifically developed for Britain's first nuclear submarine, HMS *Dreadnought*. It was so coveted that the French, unwilling to buy a foreign sonar, asked the British for the 2001 production drawings. The British refused.[5] Up to January 1961 Portland was carrying out other important research work for *Dreadnought*: for its torpedoes; on how to withstand the greater diving depths expected of a nuclear submarine; and on a new on-board computer digital weapons control system. There was separate and important work too proceeding on torpedoes, and surface sonars. With Molody directing the Portland spies towards intelligence of this type, which was of particular value to the Russians in devising counter-measures, it is highly likely that Houghton and Gee stole details of some of these research projects.[6]

Intelligence from Portland about Type 2001 and surface ship sonar undoubtedly assisted USSR research and development of more powerful underwater detection equipment and more silent submarines. These improvements took several years to effect, because the Soviet underwater acoustic research effort was compartmentalised, over-staffed and inefficient.[7] From drawings of the array of the 2001 sonar – even very general ones – Russian experts could have made good predictions about its performance.[8]

Around 1967 the first of the second generation of Soviet nuclear and non-nuclear submarines were launched, including the Yankee-class nuclear-propelled ballistic missile submarines, and Victor- and Charlie-class attack submarines. When they first encountered these vessels, British nuclear submariners were intrigued and then astounded. The characteristics of their active sonars were remarkably similar to those of the Type 2001, and in particular the 2001 steerable beam of sonar (both active and passive) called 'Sector'.[9] This confirmed what many in the Admiralty had suspected for several years: that the Portland traitors had passed secrets of the Type 2001 to the Russians, who had copied significant parts of the British sonar, and especially its active component. NATO called the new Soviet sonar 'Shark Teeth', and some British submariners, because of its striking similarity to their system which inspired it, gave it the nickname

'2001sky'. As the years passed, the West learnt more about how the Soviets had exploited sonar secrets stolen from Portland.[10]

By 1975, the West's lead over the Soviet Union in submarine sonar technology had been severely eroded. Early that year Britain's Joint Intelligence Committee was informed that: 'as a result of major improvements ... [the Russians] are now probably capable of trailing many Western submarines by using active sonar. But because of continuing deficiencies in passive sonar and noise reduction, they are still relatively vulnerable to counter-detection by many Western submarines.' One important reason why the West stayed ahead of Soviet sonar was the human factor. Early nuclear submarine sonars depended on skilful interpretation of visual and aural information available to the crews, and the defection in the 1990s of a small number of Russian submarine weapons specialists confirmed how much less efficiently Soviet submarines operated compared to their Western counterparts: drunkenness on board, for example, was widespread.[11]

The former senior KGB officer Vasili Dozhdalev, who handled Houghton before he was transferred to Molody, later confirmed: 'The advantages the Soviet navy got from the Portland case – I can draw a comparison between two competing companies, when one company steals technology from another company and can produce a competing product.' One Russian source suggested in 2013 that Molody's Portland espionage saved the USSR around fifteen billion US dollars in research costs. This undoubted hyperbole is impossible to check, but does underline the overall value the Soviets extracted from the Portland intelligence.[12] Part of that value resulted from the material including, indirectly, classified information about 'Dimus' sonar, which had originated in the USA. In assuring President Kennedy that no American secrets had been compromised, Harold Macmillan was perhaps being a little parsimonious with the truth.

It is important to place in perspective the harm caused by the Portland espionage network. The infamous Walker spy ring passed crucial US navy secrets, including cryptographic information, to the KGB for eighteen years from 1967. The treachery of John Walker and his accomplices, in the words of the Director of US Naval Intelligence, possessed the 'potential – had conflict erupted between the two superpowers – to have powerful war-winning implications for the Soviet side'.[13] The intelligence the Russians gleaned from Portland did not have such potentially severe consequences. Despite various improvements to their submarine sonars during the Cold War, the Soviet Union never caught up with the West. British and US submariners remained confident they had the edge over the Russians because their submarines were quieter, better maintained and operated, and all their submariners were volunteers and better trained. Although the passing of secret intelligence about British sonar developments from Portland to the USSR undoubtedly wreaked grave damage on Western naval interests, the evidence shows it had no long-term effect on the balance of power between NATO and Warsaw Pact forces.[14] The damage caused by the other espionage activities of the network – especially the clandestine material purloined from Porton Down and by the unidentified Agent K – is impossible to gauge without more, and solid, information from the KGB archives. But evidence points to Molody's ring of spies extending beyond Houghton and Gee (and, fleetingly, Melita Norwood).

On 28 June 2017, the section of the SVR which manages illegal spies celebrated its ninety-fifth anniversary. To mark the event Vladimir Putin gave an interview to the main state television channel in which he confirmed that he himself had worked in Directorate S (not as an agent, but in the legal *rezidentura* responsible for communication with illegals), and lavished praise on Russian intelligence illegals as 'people of special qualities', because they were ready to 'give up their life, their nearest and dearest and leave the country for many years, and to dedicate one's life to the Fatherland'. Putin attended a private celebration of the same anniversary at which he gave a congratulatory speech: 'I want to underline that our country . . . must know that among the Russian secret service served – and are still serving – real fighters. They are honest and brave professionals. We are proud of them . . . The history of the illegal secret service was created by legends . . . among them . . . Konon Molody.'[15] The following year, to coincide with a photographic exhibition, the SVR published a glossy brochure printed in gold with a foreword by its director, Sergey Naryshkin, featuring portraits of a selection of Russia's 'golden spies', placing Lona and Morris Cohen (each with their own page) in the company of such remarkable agents as Willie Fisher, Kim Philby and George Blake. This recognition of Molody and the Cohens by the Russian state and intelligence services echoes that of earlier years: the medals, the laudatory articles and programmes in the media and the postage stamps. The propaganda message is clear: the illegals department of Russian intelligence is in rude health in the first quarter of the twenty-first century with a history to celebrate stretching back almost a century, and standing high in its pantheon of heroes are Konon Molody and Morris and Lona Cohen.

For Western intelligence services, the Portland spy ring became a historical curiosity. It echoed for a decade or two in the collective memory of the Security Service while the Cold War exercised its chilling grip. When Stella Rimington was working in MI5 counter-espionage in the early 1970s, the renegade officer Peter Wright was referred to by the nickname of the 'KGB illegal', because 'with his appearance and his lisp we could imagine that he was really a KGB officer himself, living under a false identity, perhaps like Gordon Lonsdale'.[16]

David Whyte, the MI5 officer who started the hunt for the Portland spies, went on to enjoy a distinguished career in the Security Service. He was promoted to be Director of F Branch (counter-subversion – home) and C Branch (protective security) before officially retiring in 1975. Whyte continued to work part-time for MI5, writing an internal post-war history of the Security Service and helping with recruitment interviews until 1980. With his health declining, he moved in 1991 to Aldeburgh on the Suffolk coast, popular with a number of active and retired British intelligence officers attracted by its isolation and stark beauty, but chosen by Whyte because of happy memories of childhood and later holidays. He died in 2005.[17]

For Charles Elwell the investigation of the Portland spies was in some ways the apogee of his MI5 career. He was immensely proud to be invited to Buckingham Palace in the summer of 1961 to receive an OBE from the Queen. He was in charge of the successful John Vassall investigation the following year, before being seconded to the Ministry of Defence in 1965 to improve security. He returned to MI5's counter-espionage branch three years later and in 1974 switched to counter-subversion. Here, the characteristic that made Elwell such a superlative investigator of Soviet spies – single-minded focus once his eye was on a goal he believed in – was regarded by some colleagues as a hindrance. Elwell hated dissembling, and regarded a number of more extreme left-wingers as subversives, furtively concealing their allegiance to Soviet-led international communism. A number of his senior Security Service colleagues considered he was over-inclined to spot subversion where none existed. This, allied to his direct manner, his (in Elwell's own words) 'unfortunate capacity for being a nuisance', lack of flexibility when his mind was made up, and as he saw it an unwillingness to hobnob with colleagues to play office politics, blocked him from further promotion, and he retired, frustrated, in 1979. After an active retirement of fruit-farming, and writing about left-wing infiltration both real and imagined, he died in 2008.[18]

Soon after the Portland spy trial in Washington in June 1961, Alan Belmont was promoted to Assistant to the Director in charge of all investigative work of the FBI, the number-three position in the Bureau. In this role he spearheaded the investigation into the assassination of President John Kennedy (being accused by some of involvement with a cover-up), and was involved in helping Hoover understand the disgraceful attempts of William C. Sullivan, who took over his role as head of the Domestic Intelligence Division, to smear Martin Luther King. He retired from this bruising role in December 1965, and then worked for the Hoover Institution for a period. He died in August 1977 as he had lived his years in the FBI, believing, as he argued in a summer lecture he gave after retirement, that the Bureau had a central role to play in the 'preservation of American freedom and democracy'.[19]

By the time Jonathan Evans joined MI5 in 1980 (he became Director General in 2007), the Portland spy ring was mentioned briefly during initial training, but within a decade or two Konon Molody and the Cohens had faded into espionage ghosts from a past era. It was different behind the Iron Curtain. For two decades or so after their arrests in January 1961, the work of Molody and the Cohens was taught to new KGB officers as a training case – how KGB illegals can create and run a network of agents and their means and methods – even if in Directorate S by the 1980s new trainee illegals knew who they were and what they had done in vague terms but little detail. The memory of Molody and the Cohens was also kept alive by a steady flow of laudatory media articles and books, even including in 2001 two leaden volumes of prison correspondence of the Cohens (discovered in a Peek Freans biscuit tin in Morris Cohen's Moscow flat by KGB colleagues after his death). For Russian intelligence and

Putin's Russia, these three agents are not actors in some historical espionage drama but have resonance today, as illustrious examples of sacrifice and spycraft.

They were certainly among the greatest KGB illegals who operated during and after the Second World War. Each worked for Moscow Centre for a decade or so before being arrested. All three were swapped and released from prison, in each case with Russia reaping the advantage from the exchange of pieces on the international espionage chessboard, demonstrating to other KGB agents that Moscow would look after its own. The three spies successfully played their allocated parts, whether it be (like the Cohens) as couriers, recruiters or communications operators, or (like Molody) becoming the KGB's first illegal *rezident* in Britain of the post-war years. Their achievements as intelligence agents are all the more remarkable because of their rarity. A number of KGB post-war illegals sent to North America in the 1950s, for example, were failures, succumbing to loneliness, alcohol, boredom or the temptations of the West. Many others were deployed not against the West but against dissidents in the Soviet Bloc, especially in the 1960s. None, however, matched the outstanding record of the great illegals of the 1930s, like Arnold Deutsch, who recruited Kim Philby and the other four of the 'Magnificent Five' Cambridge spies.[20]

Russian illegals still pose a clear and present threat to the West. This was evident in 2010, when ten of them were arrested in the USA by the FBI after an investigation known as 'Operation Ghost Stories', but were exchanged the following month for several individuals – including Sergey Skripal of Russia's military intelligence agency, the GRU – who had been jailed in Russia for spying for the West. Significantly, the ring had initially been betrayed by a Russian defector to the USA, Alexander Poteyev, who had enabled the FBI to place the network under surveillance. The lesson once again for Russia was that its illegals are most vulnerable to betrayal by agents on their own side – as Molody and the Cohens had been betrayed by Goleniewski fifty years before.[21]

Of course, in the digital age of the twenty-first century Department S of the SVR faces challenges completely undreamt of when Molody and the Cohens operated as illegals. Information about individuals can be harvested and compared more effectively and swiftly than before, making the creation and maintenance of 'legends' for illegals in some ways more demanding than ever, but in others offering new possibilities and horizons. Digital information can, however, be digitally manipulated, and the potential prize of a spy being able to work with less risk of surveillance is so enticing that Russia continues to operate an illegals programme worldwide. The Russians are not alone. The intelligence services of China, Israel, East European countries and Cuba have all run illegals in the recent past and are almost certainly doing so now. By contrast the CIA, MI6 and Western services have traditionally preferred to cultivate contacts to provide 'non-official covers' (NOCs), whether in journalism or business, as a pretext under which an officer can operate, or to recruit their own or foreign citizens to carry out tasks on their behalf. The CIA dabbled with illegals in trying to penetrate al-Qaeda and ISIS, but concluded that the benefits did not outweigh the risks.[22]

The interior of the Cohens' bungalow in Ruislip, 45 Cranley Drive, has been modernised since they were arrested on 7 January 1961. Still recognisable, however, is the exterior, and in the kitchen the trapdoor could until recently be heaved up to reveal the cellar below, partly excavated by the KGB spies in the 1950s. The current owner, a London Tube driver, occasionally hears a tentative knock on the door from curious espionage voyeurs. One of the most re-markable revealed a smiling group of Chinese people who said, not very convincingly, that they were merely tourists. With the Chinese intelligence service known to be one that uses illegals abroad to gather secret intelligence, it may not be fanciful to wonder if they were an advance guard of trainee illegals visiting a shrine to their future profession.[23]

Afterword

After publication of *Dead Doubles* I received an intriguing message from the granddaughter of Harry Houghton. 'Harry', the message said, 'was no angel but he was no wife-beater.' According to Houghton's daughter and granddaughter, 'Peggy' Johnson was not the downtrodden character she presented herself to be to the Security Service. Rather, she was a resilient and sly woman, with a natural actor's gift to manipulate those around her for advantage. Born Amy Smith in 1895 in Chorley, Manchester, into tough circumstances, she had been married to two men before meeting Harry Houghton around 1929. Harry brought up Amy's young daughter by a previous marriage as his own. According to her, Harry was no drunken wife abuser. He was at heart a kind man, indeed more empathetic in some ways than her own mother.[1]

Documents released by the FBI confirm that the Bureau was still having the fingerprints of the Cohens checked against police files around the world as late as December 1960 – only days before their arrest in London. The spy case was important enough in America for FBI Director J. Edgar Hoover to brief at the start of February 1961 not only President Kennedy, but his brother, Robert, who was Attorney General, and the Secretary of State Dean Rusk. Hoover told them that the Cohens' re-emergence in England as KGB illegal spies in 1954, with new identities after vanishing from New York without a trace four years before, was 'an outstanding example of the long-range planning done by Soviet intelligence.'[2]

Intriguing new titbits of information about the Soviet agents Lona and Morris Cohen and Konon Molody continue to emerge in Russia. In summarising the careers of these leading illegals, the SVR (Russia's main foreign intelligence agency) confirmed for the first time that Lona Cohen had in 1948–50 'ensured the transfer to Moscow of important documentary materials on the creation of the atomic bomb in the United States' and that during these years Lona and Morris Cohen acted as the liaison and couriers between Rudolf Abel/Willie Fisher and various Soviet agents 'working in nuclear facilities.'[3]

TB
London, June 2021

Acknowledgements

Any writer of a non-fiction book which straddles journalism and history, contemporary sources and archives, knows that although there may be just the author's name on the cover, their book is only possible because of the help and contributions of many others. This is especially true of espionage, where the intelligence services of various nations and the individuals who have worked for them understandably guard their secrets tightly.

Many individuals have contributed to this history. As regards the UK, I would first like to thank Christopher Andrew. Doyen of intelligence historians, mentor and guide, Chris has in some ways been the godfather to this book, and unstinting in providing support and expertise throughout its creation. From first contacting them, the four children of Charles Elwell (Charles, Selina, Henry and Henrietta) have been most generous with their time and help. Similarly, the two children of David Whyte, Jocelyn Stell and Adrian Whyte, could not have given me more assistance. Michael Bonkowski generously tracked down and translated various sources in Polish about Goleniewski. In America, David Robarge of the CIA, John Fox of the FBI and David Rosenberg of the Institute for Defense Analyses were endlessly patient and creative in answering my questions; and I am most grateful for the excellent help in Moscow afforded to me by a handful of former officers of the Russian Intelligence Service, my main researcher, Sofi Kolomiets, and Alexey Nikolov. Various former officers of MI5 have assisted me in various ways on the condition of anonymity: their contribution is no less valued although their names remain in the shadows. I wish to thank everyone who gave me the interviews referred to in the Notes, whether in Britain, the United States or Russia, in many cases kindly finding time to answer follow-up questions. Their names are listed under Sources, but simply for reasons of space I have not also listed them here.

Others have assisted me in various ways too complicated or

numerous to explain, but without their help writing parts of this book would have been either impossible, delayed or considerably less pleasurable or interesting. I wish to thank most warmly: Mark and Diane Abbott, Richard Aldrich, Witold Bagieński, Gill Bennett, Nigel Chapman, Sheila Collins, Kate Easton, David Gioe, Michael Goodman, John Hattendorf, Carenza Hayhoe, Tom Hayhoe, Peter Hennessy, Carolyn Ireland, James Jinks, Carolyn and Mike Jones, Fiona Jones, Gary Kern, Svetlana Lokhova, Andrew Lownie, Tom Maguire, Stuart Morris, David Omand, Hayden Peake, Evgeniya Popova, the Portland Museum, Nicki Quinn, Nicholas Reynolds, Phylly and Geoff Shepherd, Sean Stevens, Jan Tipler, Mark Urban, Andrey Vedyaev, Calder Walton and Nigel West (Rupert Allason).

My publishers – Alan Samson, and his team at Weidenfeld, and Jennifer Barth and her stalwarts at HarperCollins USA – have been a source of constant support and creative input, as have my agent, Bill Hamilton, and his assistant, Florence Rees. Particular thanks also to Ray Batvinis, Hayden Peake, Peter Molloy and Fran O'Brien, who found time to comment incisively on my manuscript; and to David Cust and John Edgell (both former commanders of British nuclear submarines) for reviewing my assessment of the damage to the West's maritime interests caused by the espionage of the Portland ring.

Finally I wish to thank my wife, Sally Gaminara. From helping to craft the modest proposal to publishers to the final book, she has been outstandingly helpful and patient. My thanks and love also to my remarkable children, Katherine and George, whose lives have been affected by the writing of this book in ways which only they and my wife fully appreciate.

Sources

Manuscript sources

Diaries of Charles Lister Elwell, in possession of his elder son (Charles Elwell)

Harold Macmillan diaries, Bodleian Library, Oxford University

Alexander Vassiliev Notebooks, Wilson Center Digital Archive, Library of Congress: available at
https://digitalarchive.wilsoncenter.org/collection/86/vassiliev-notebooks

Papers of David Whyte, in possession of his daughter (Jocelyn Stell)

Archives

Papers of Morris Cohen (aka Peter Kroger), Box No: 06/80/1, Imperial War Museum, London

John F. Kennedy Presidential Library, Boston, Robert F. Kennedy Attorney-General Papers

The National Archives (TNA), Kew, London (all ADM [Admiralty], CAB [Cabinet], CRIM [Criminal Courts], KV [MI5], and PREM [Prime Minister] files referred to in the Notes are stored here)

Author's interviews (all interviews took place in the UK unless otherwise stated)

Vladimir Chikov, Moscow
David Cust
Nikolay Dolgopolov, Moscow
Kate Dumville
John Edgell
Charles Elwell, Madrid
Henry Elwell
Selina Elwell
Jonathan Evans (conversation after a public event, London, 4 July 2018)
Michael Herman
James Jinks
Vitali Korotkov, Moscow

Mikhail Lyubimov, Moscow
Caroline Macmillan
Eliza Manningham-Buller
Donald Nairn
NATO source (from a NATO country with contacts in the Russian intel-
    ligence services)
Val Parker
David Parry
Nathan Prince
David Robarge, Washington DC
John Sipher, Washington DC
Jocelyn Stell
Nigel West (real name Rupert Allason)

Documentaries and broadcasts

BBC *Southern Eye* documentary, 'Portland Spy Ring'. Broadcast in 1993.
    Available at: https://www.youtube.com/watch?v=Puw3OWsnQlI&-
    feature=youtu.be
    Accessed: June 2019
'The Secret Underworld', created/produced September 1963, US Defense
    Department, US National Archives (College Park), Local Identifier
    330.59

Secondary sources: published books, memoirs, published docu-
ments and reports

Agranovsky, Valery, *Razvedchik 'Mertvogo sezona'* [The Spy of 'Dead
    Season'] (Moscow: Algoritm, 2008)
Albright, Joseph and Kunstel, Marcia, *Bombshell: The Secret Story
    of America's Unknown Atomic Spy Conspiracy* (New York: Times
    Books, 1997)
Aldrich, Richard J., *The Hidden Hand: Britain, America and Cold War
    Secret Intelligence* (London: John Murray, 2001)
Aldrich, Richard J., *GCHQ: The Uncensored Story of Britain's Most
    Secret Intelligence Agency* (London: Harper Press, 2010)
Aldrich, Richard and Cormac, Rory, *The Black Door: Spies, Secret
    Intelligence and British Prime Ministers* (London: William Collins,
    2016)
Andrew, Christopher, *The Defence of the Realm* (London: Penguin
    Books, 2010)
Andrew, Christopher, *The Secret World: A History of Intelligence* (Lon-
    don: Allen Lane, 2018)

Andrew, Christopher and Gordievsky, Oleg, *Instructions from the Centre: Top Secret Files on KGB Foreign Operations 1975–1985* (London: Sceptre, 1993)

Andrew, Christopher and Gordievsky, Oleg, *KGB: The Inside Story of its Foreign Operations from Lenin to Gorbachev* (London: Hodder & Stoughton, 1990)

Andrew, Christopher and Mitrokhin, Vasili, *The Mitrokhin Archive: The KGB in Europe and the West Defence of the Realm* (London: Penguin Books, 1999)

Antonov, Vladimir S., *Vokopakh kholodnoy voiny* ['In the Trenches of the Cold War'] (Moscow: Veche, 2010)

Antonov, Vladimir S., *Konon Molody* (Moscow: Molodaya Gvardiya, 2018)

Antonov, Vladimir S. and Karpov, V.N., *Nelegalnaya razvedka* ['Illegal Spies'] (Moscow: Mezhdunarodnye otnosheniya, 2013)

Arthey, Vin, *Abel: The True Story of the Spy They Traded for Gary Powers* (London: Biteback Publishing, 2015)

Bagley, Tennent H., *Spy Wars: Moles, Mysteries, and Deadly Games* (New Haven and London: Yale University Press, 2007)

Barsky, Jack, *Deep Undercover: My Secret Life* (New York: Tyndale Momentum, 2017)

Bingham, Charlotte, *MI5 and Me: A Coronet Among the Spooks* (London: Bloomsbury, 2018)

Blake, George, *No Other Choice* (London: Jonathan Cape, 1990)

Bower, Tom, *The Perfect English Spy* (London: Heinemann, 1995)

Bulloch, John and Miller, Henry, *Spy Ring: The Full Story of the Naval Secrets Case* (London: Secker and Warburg, 1961)

Burke, David, *The Spy Who Came in from the Co-op: Melita Norwood and the Ending of Cold War Espionage* (Woodbridge and Rochester, England: Boydell Press, 2008)

Carr, Barnes, *Operation Whisper: The Capture of Soviet Spies Morris and Lona Cohen* (Lebanon, New Hampshire: ForeEdge, 2016)

Carrington, Lord, *Reflect on Things Past: The Memoirs of Lord Carrington* (London: Collins, 1988)

Carter, G.B., *Chemical and Biological Defence at Porton Down 1916–2000* (London: The Stationery Office, 2000)

Catterall, Peter (ed.), *The Macmillan Diaries,* vol. II, *Prime Minister and After, 1957–1966* (London: Macmillan, 2011)

Cherkashin, Victor, *Spy Handler: Memoir of a KGB Officer* (New York: Basic Books, 2005)

Chikov, Vladimir M., *Sekretniye missii: Nelegaly chast 2, 'Dachniki' v Londone* ['Secret Missions: Illegals Part 2, "Dachniki" in London'] (Moscow: Terra, 2001)

Close, Frank, *Half-Life: The Divided Life of Bruno Pontecorvo, Physicist or Spy* (London: Oneworld Publications, 2015)

Close, Frank, *Trinity: The Treachery and Pursuit of the Most Dangerous Spy in History* (London: Penguin Random House, 2019)

Corera, Gordon, *Russians Among Us: Sleeper Cells, Ghost Stories and the Hunt for Putin's Agents* (London: William Collins, 2020)

Davenport-Hines, Rupert, *An English Affair: Sex, Class and Power in the Age of Profumo* (London: William Collins, 2013)

Deloach, Cartha 'Deke', *Hoover's FBI* (Washington DC: Regnery Publishing, 1997)

Dolgopolov, Nikolay, *Glavny protivnik: Tainaya voina za SSSR* ['The Main Enemy: The USSR's Secret War'] (Moscow: Eksmo, 2011)

Dolgopolov, Nikolay, *Legendarnye razvedchiki. Na peredovoy vdali ot fronta – vneshniaya razvedka v gody Velikoy Otechestvennoy voiny* ['Legendary Spies: On the Frontline far from the Front – the External Intelligence Service during the Years of the Great Patriotic War'] (Moscow: Molodiya gvardiya, 2018)

Drozhdov, Yuri I., *Nuzhnaya rabota: zapiski razvedchika* ['Necessary Job: Notes of an Intelligence Officer'] (Moscow: VlaDar, 2004)

Elwell, Charles, *An Autobiography* (2005: private circulation)

Evseev, A.E., with Gubernatorov, N.V., Korneshov, L.K. and Molodaya, G.P., *Moya professiya – razvedchik (vospominaniya ofitsera KGB* ['Intelligence Is my Profession' (Memoirs of a KGB Officer)] (Moscow: Orbita, 1990)

Feklisov, Alexander, *The Man Behind the Rosenbergs* (New York: Enigma, 2001)

Gaddis, John L., *The Cold War* (London: Allen Lane, 2005)

Gromushkin, Pavel G., *Razvedka: Ludi, portrety, sud'by* ['Intelligence Service: People, Portraits, Destinies'] (Moscow: Dobrosvet-2000, 2002)

Haynes, John Earl, with Klehr, Harvey and Vassiliev, Alexander, *Spies: The Rise and Fall of the KGB in America* (New Haven and London: Yale University Press, 2009)

Hennessy, Peter, *The Prime Minister: The Office and its Holders since 1945* (London: Allen Lane, 2000)

Hennessy, Peter, *The Secret State: Preparing for the Worst 1945–2010* (London: Penguin Books, 2010)

Hennessy, Peter, *Winds of Change: Britain in the Early Sixties* (London: Allen Lane, 2019)

Hennessy, Peter and Jinks, James, *The Silent Deep: The Royal Navy Submarine Service since 1945* (London: Penguin, 2016)

Hermiston, Roger, *Greatest Traitor: the Secret Lives of Agent George Blake* (London: Aurum Press, 2014)

Horne, Alistair, *Macmillan 1957–1986*: vol. 2 of the *Official Biography* (London: Macmillan, 1989)

Houghton, Harry, *Operation Portland: The Autobiography of a Spy* (London: Rupert Hart-Davis, 1972)

Kolosov, Leonid and Molody, Trofim, *Mertvy sezon, Konec legendy* ['Dead Season: End of the Legend'] (Moscow: Kollektsiya Sovershenno sekretno, 1998)

Kouzminov, Alexander, *Biological Espionage* (London: Greenhill Books, and Pennsylvania: Stackpole Books, 2005)

Kroger, Peter and Helen, with Lonsdale, Gordon, *Pisma iz turem eje velichestva 1961–9* ['Letters from Her Majesty's Prisons'], 2 vols. (Moscow: Zhilin, 2001)

Kynaston, David, *Modernity Britain 1957–62* (London: Bloomsbury Paperbacks, 2015)

Lamphere, Robert J. and Shachtman, Tom, *The FBI–KGB War: A Special Agent's Story* (London: W.H. Allen, 1987)

Lewis, Norman, *Spycatcher: A Biography of Detective-Superintendent George Gordon Smith* (London: W.H. Allen, 1973)

Lonsdale, Gordon, *Spy: Twenty Years of Secret Service* (London: Neville Spearman, 1965)

Macintyre, Ben, *The Spy and the Traitor: The Greatest Espionage Story of the Cold War* (London: Viking, 2018)

Macmillan, Harold, *At the End of the Day 1961–1963* (London: Macmillan, 1973)

Molodaya, Natalia, *Istoriya semyi Molodyh* ['History of the Molody Family'] (Moscow: Mtusi, 2007)

Murphy, David E., Kondrashev, Sergei A. and Bailey, George, *Battleground Berlin: CIA vs. KGB in the Cold War* (New Haven: Yale University Press, 1997)

Nicholson, Virginia, *How Was It For You?: Women, Sex, Love and Power in the 1960s* (London: Viking, 2019)

Omand, David (entry on Arthur Bonsall), *Dictionary of National Biography* (Oxford: OUP, 2018)

Pavlov, Vitaly, *Operatsiya Sneg* ['Operation "Snow"'] (Moscow: Geya, 1996)

Pavlov, Vitaly, *Tragedii sovetskoy razvedki* ['Soviet Spy Tragedies'](Moscow: Tsentrpoligraf, 2000)

Pawlikowicz, Leszek, *Tajny front zimnej wojny. Uciekinierzy z polskich służb specjalnych 1956–1964*,['The Secret Front in the Cold War. Defectors from Polish Secret Services 1956–1964'] (Warsaw: RYTM, 2004)

Polmar, Norman and Moore, Kenneth, *Cold War Submarines: The Design and Construction of US and Soviet Submarines* (Washington DC: Potomac Books, 2004)

Polmar, Norman and Whitman, Edward, *Hunters and Killers,* vol. 2: *Anti-Submarine Warfare from 1943* (Annapolis, Maryland: Naval Institute Press, 2016)

Powers, Richard Gid, *Broken: The Troubled Past and Uncertain Future of the FBI* (New York: Free Press, 2004)

Primakov, Evgenii et al., *Ocherki istorii rossiiskoy vneshney razvedki* ['Essays on the History of Russian International Intelligence'], vol. 5, *1945–1965* (Moscow: Mezhdunarodnye, 2003)

Prokopenko, Igor, *Yaderny schit Rossii: kto pobedit v 3 mirovoy voine* ['The nuclear shield of Russia: who will win in World War Three'] (Moscow: Eksmo, 2016)

Rimington, Stella, *Open Secret: The Autobiography of the Former Director-General of MI5* (London: Arrow, 2002)

Ring, J., *We Come Unseen: The Untold Story of Britain's Cold War Submariners* (London: John Murray, 2001)

Robarge, David, *John McCone: As Director of Central Intelligence 1961–1965* (Washington: Center for the Study of Intelligence). Declassified, 10 October 2015. Available at: https://www.cia.gov/library/readingroom/collection/john-mccone-director-central-intelligence-1961-1965

Sampson, Anthony, *Anatomy of Britain* (London: Hodder and Stoughton, 1962)

Sampson, Anthony, *Anatomy of Britain Today* (London: Hodder and Stoughton, 1965)

Sandbrook, Dominic, *Never Had It So Good: A History of Britain from Suez to the Beatles* (London: Abacus, 2006)

Schmidt, Ulf, *Secret Science: a Century of Poison Warfare and Human Experiments* (Oxford: Oxford University Press, 2015)

Serov, Ivan, *Zapiski iz chemodana. Tainye dnevniki pervogo predseda telya KGB* ['Notes from a Suitcase: Secret Diaries of the KGB's First Chairman'] (Moscow: Abris, 2017)

Shackley, Ted, with Finney, Richard A., *Spymaster: My Life in the CIA* (Dulles, Virginia: Potomac Books, 2005)

Sisman, Adam, *John Le Carré: The Biography* (London: Bloomsbury, 2015)

Snelling, O.F., *Rare Books and Rarer People: Some Personal Reminiscences of 'The Trade'* (London: Werner Shaw, 1982)

Sulick, Michael J., *American Spies: Espionage against the United States from the Cold War to the Present* (Washington DC: Georgetown University Press, 2013)

Tchikov, Vladimir, with Kern, Gary, *Comment Staline a volé la bombe atomique aux Américains: dossier KGB no.13676* ['How Stalin Stole the Atomic Bomb from the Americans: KGB file no.13676'](Paris: Robert Laffont, 1996)

Thorpe, D.R., *Supermac: The Life of Harold Macmillan* (London: Chatto & Windus, 2010)

Tietjen, Arthur, *Soviet Spy Ring* (London: Pan Books Limited, 1961)

Tucker, Jonathan, *War of Nerves: Chemical Warfare from World War I to Al-Qaeda* (New York: Pantheon, 2006)

Tuorinsky, Shirley D. (general editor), *Medical Aspects of Chemical Warfare* (Washington DC: Office of Surgeon General, 2008)

Urban, Mark, *The Skripal Files: The Life and Near Death of a Russian Spy* (London: Macmillan, 2018)

Vitkovsky, Alexander, *Poedinok spetssluzhb* ['The Duel between Intelligence Services'] (Moscow: Veche, 2013)

Weinstein, Allen and Vassiliev, Alexander, *The Haunted Wood: Soviet Espionage in America – the Stalin Era* (New York: Random House, 1999)

West, Nigel, *The Illegals* (London: Hodder & Stoughton, 1993)

West, Nigel and Tsarev, Oleg, *Crown Jewels: The British Secrets at the Heart of the KGB Archives* (London: HarperCollins, 1998)

Whitemore, Hugh, *Pack of Lies* (London: Samuel French, 1983)

Whittell, Giles, *A True Story of the Cold War: Bridge of Spies* (London: Simon & Schuster, 2010)

Wise, David, *Molehunt: The Secret Search for the Traitors that Shattered the CIA* (New York: Random House, 1992)

The WPA Guide to New York City (originally published in 1939, reprinted in New York: Random House, 1992)

Wright, Peter, *Spycatcher* (Victoria, Australia: William Heinemann, 1987)

Yakunin, Vladimir, *The Treacherous Path: An Insider's Account of Modern Russia* (London: Biteback, 2018)

Secondary sources: published chapters, journal articles and papers

Aldrich, Richard J., 'British Intelligence and the Anglo-American "Special Relationship" during the Cold War', *Review of International Studies*, 24 (3) (July 1998), p. 331

Bagieński, Witold, 'Analiza sprawy ppłk. Michała Goleniewskiego, uciekiniera z wywiadu PRL'['Analysis of the Case of Lt-Col. Michał Goleniewski, defector from the intelligence service of Communist Poland'], *Studia nad wywiadem i kontrwywiadem Polski w XX wieku*, tom 3, pod red. W. Skóry i P. Skubisza (Szczecin: IPN, 2016)

Carter, Gradon and Balmer, Brian, 'Chemical and Biological Warfare and Defence, 1945–90', in Robert Bud and Philip Gummett (eds), *Cold War, Hot Science: Applied Research in Britain's Defence Laboratories, 1945–1990* (London: Science Museum, 2002), pp. 295–338

Elwell, Charles, under the pseudonym 'Elton', 'A Russian Intelligence Officer Exposed: Konon Trofimovich Molody' by An Officer of the Security Service, *Police Journal*, vol. XLIV, no. 2 (April–June 1971), p. 111

Hewitt, Steve, '"Strangely Easy to Obtain": Canadian Passport Security, 1933–73', *Intelligence and National Security*, vol. 23, no. 3 (June 2008), p. 381

Parry, David, 'The History of British Submarine Sonars', 2018, at rnsubs.co.uk/articles/development/sonar.html (accessed June 2019)

Platt, Roger H., 'The Soviet Imprisonment of Gerald Brooke and Subsequent Exchange for the Krogers, 1965–1969, *Contemporary British History*, vol. 24, no.2 (June 2010), p. 193

Preston, John, 'The Influence of the Cold War on Submarine Design', in Martin Edmonds (ed.), *100 Years of the Trade: Royal Navy Submarines Past, Present & Future* (Lancaster, UK: Centre for Defence and International Security Studies, 2001), pp. 74–80

Smith, George G., 'Soviet Spy Ring', *Police College Magazine*, Autumn 1962, p. 288

Wright, Tom, 'Aircraft Carriers and Submarines: Naval R&D in Britain in the Mid-Cold War', in Robert Bud and Philip Gummett (eds), *Cold War, Hot Science: Applied Research in Britain's Defence Laboratories, 1945–1990* (London: Science Museum, 2002), pp. 147–83

Secondary source: unpublished memoir

Belmont, Alan H., 'As I recall it! Incidents in the life of a G-man', unpublished manuscript, date unclear but after Belmont's retirement in 1965, courtesy of John Fox, FBI

Notes

All short-title references in the Notes section can be found in full in the Sources.

Epigraph

1 *Thoughts and Adventures* (London: Odhams Press Ltd, 1949), pp. 58–9.
2 *Selected Poems*, translated by David McDuff (Newcastle upon Tyne: Bloodaxe Books, 1991), p. 102.

Prologue

1 Although mainly based on documents in KV 2/4430 and 4431 (all KV 2 files for MI5 are in TNA), this account of the 12 September 1960 MI5 search also relies on: Wright, *Spycatcher*, pp. 39–40, 130–31; Andrew, *Defence of the Realm*, p. 486; and the Elwell MI5 'biography' of Lonsdale dated 18 October 1960 in KV2/4383.
2 Elwell note for file, 14 September 1961: KV 2/4430.
3 Made of highly inflammable cellulose nitrate impregnated with a zinc compound, usually zinc oxide.

Part One

1 The account of the MI5 Portland spy ring investigation for the period 15 December 1954 to 23 June 1960 in Part One is based on documents in Houghton file KV 2/4380, unless otherwise stated.
2 See Roger Hollis evidence to Romer Inquiry, 18 April 1961, pp. 3–6: CAB 301/252.
3 Interviews, Caroline Macmillan and Val Parker, MI5 secretaries, recruited in 1960 and 1961 respectively; Bingham, *MI5*, *passim*; Andrew, *Defence of the Realm*, pp. 336–8.
4 Stell, interviews. Whyte, unpublished notes on his Second World War experiences: Whyte papers.

5 For the history of 'Sniper' see: Wright, *Spycatcher*, pp. 128–9; Shackley and Finney, *Spymaster*, pp. 26–7; Bagley, *Spy Wars*, pp. 48–9, 61, 280–81; Bower, *Perfect English Spy*, pp. 257–8; Murphy et al., *Battleground*, p. 497, fn. 4; Robarge interview. The MI5 code-name for 'Sniper' was 'Lavinia', but in this book for clarity I refer throughout to the agent as 'Sniper'.

6 In the summer of 1958, for example, an MI6 source had reported that around 1953 the Polish intelligence service had recruited one of the assistant Service attachés at the British embassy in Warsaw: p. 2 of undated note attached to letter, Hollis to Brook, 22 March 1961: CAB 301/249.

7 Leggett died in 2012. See Andrew, *Defence of the Realm*, p. 328; Wright, *Spycatcher*, pp. 320–21, where Wright refers to Leggett as 'Gregory Stevens'.

8 Private information. James Craggs is a pseudonym. The name of the case officer is redacted from the released MI5 files. The author discovered his real identity but was requested by MI5 sources not to name him to avoid potential distress to his family.

9 Confirmed by MI5's Director General himself, Roger Hollis, who said KGB officers were highly trained 'for the very purpose of fooling us', including in the use of dead letter boxes. 'The popular one is the public lavatory where they are alone, and you can conceal the thing [information or money] behind the cistern.' Hollis evidence to Romer, 25 May 1961, p. 10: CAB 301/252.

10 Wright, *Spycatcher*, p. 44; Andrew, *Defence of the Realm*, pp. 334–5.

11 Sampson, *Anatomy of Britain*, p. 253.

12 Sampson, *Anatomy of Britain Today*, p. 347.

13 Sir Dick White, head of SIS, quoted in Bower, *Perfect English Spy*, p. 259.

14 The work involved checking each of the ninety-nine documents on the list, learning if any were missing, and investigating the security arrangements in the British embassy in Warsaw. The exercise was complicated by the fact that British intelligence were already investigating burglaries at the British consulate in Gdynia in Poland from 1955 onwards, which had resulted in the loss of various sensitive attaché documents, and reports from agents in addition to 'Sniper' of other leaked material.

15 West, *Illegals*, p. 163.

16 Hennessy, *Winds of Change*, pp. 170–75.

17 Sandbrook, *Never Had It So Good*, p. xxii.

18 Hornby to D1, 7 February 1962: KV 2/4480. Interview Pearce, 5 May 1964: KV 2/4482.

19 Wright, *Spycatcher*, p. 45.

20 The source for all MI5 documents for the period 27 June–9 August 1960 inclusive is KV 2/4381, unless otherwise stated.

21 Although most of this account of the surveillance of Houghton and Gee on 8 and 9 July is based on Skardon's note of 11 July 1960 in KV2/4381, some details are from the evidence of the two A4 watchers at the 1961 trial on 13 March: see Witnesses A and B in CRIM 1/3604, file 3.

22 MI5 Lonsdale file for 14 July–16 August 1960: KV 2/4429.

23 Stell interview; letter, Moore to Whyte, 26 July 1960: Whyte papers.

24 Higgitt to Martin, 2 August 1960; Whyte to A4, and Whyte to Higgitt, 3 August 1960: KV 2/4429.

25 Elwell obituary, *The Times*, 21 January 2008; Wright, *Spycatcher*, p. 89.

26 E.g. Elwell diary entries, 25 June and 9 July 1960: Elwell papers. Interviews with the children of Charles Elwell and David Whyte.

27 Elwell diary entry, 15 July 1960: Elwell papers. Interviews: Charles, Selina and Henry Elwell.

28 A4 surveillance report, 4 and 5 August 1960: KV 2/4429.

29 This account of what happened in the café is based on Skardon's A4 surveillance report of 8 August 1960; and evidence at the Portland trial by Witness D, 15 March 1961: file 3, CRIM 1/3604.

30 They had indeed defected and a few days later in Moscow, on 6 September 1960, gave 'perhaps the most embarrassing press conference in the history of the US intelligence community', revealing that the NSA had been decrypting the communications of key US friends and allies. See Andrew, *Secret World*, p. 675.

31 Confusingly, Lonsdale was also given the code-name of 'Trek' by Britain's Joint Intelligence Committee.

32 Notes, 8 and 12 August; A4 surveillance reports, 9 and 10 August; letter to MI6, 12 August 1960: KV 2/4429.

33 Aldrich, *GCHQ*, pp. 133–47; Polmar, *Hunters and Killers*, p. 134; Prince interview. Part of the West's secret submarine espionage programme against the USSR during the Cold War was first revealed in 1998 in *Blind Man's Bluff: the Untold Story of Cold War Submarine Espionage* by Sherry Sontag et al.

34 ASDIC was named after the Anti-Submarine Detection Investigation Committee, a body formed during the First World War in

Britain to carry out research on and experiments for the detection of submarines. SONAR (derived from SOund NAvigation and Ranging) was an American term dating from the Second World War, and is now used universally to describe all underwater detection equipment. The Royal Navy, however, with its love of tradition, continued to use the term ASDIC until around 1964 to describe systems for locating underwater objects by transmitting an acoustic pulse of energy and listening for any echoes returned from those objects.

35 As a result of Caswell's time in London he became a lifelong friend of Whyte, sending him chatty annual New Year greetings with news of mutual acquaintances in the intelligence world: Whyte papers.

36 MI5 documents relating to Lonsdale for the period 16 August–14 September 1960 are in KV 2/4430.

37 One was Kenneth Mills, MI5's representative in Tangier after the Second World War, who was later transferred to Jamaica. Whyte probably befriended him while SLO in Trinidad. See letter, Bridget Moore to her father, 13 March 1961: Whyte papers.

38 Wright, *Spycatcher*, pp. 24–5, 54; Andrew, *Defence of the Realm*, pp. 335, 486.

39 Leggett note, 14 September 1960: KV 2/4430.

40 Charles Elwell diary entries, 3 March and 5 July 1956 and 16 September 1960. Andrew, *Defence of the Realm*, pp. 486–7. Elwell appeared to attend Tom Pope's parties more out of duty than pleasure. Of a party held on 3 March 1960 he wrote, 'The party was for his fellow Chinese students at the School of Oriental Studies and their professors and lecturers. They were a depressing lot, half suburban, half Bloomsbury, lacking robustness, insipidly intellectual, underfed . . . Amongst them two F.O. men with long hair, a Communist Chinese who was abused in front of me by the attractive Canadian wife of one of the professors . . . Some of T[om]'s guests got rather drunk, one was sick in the sink and retired to T's bed, another I found sometime after midnight slumped against the door leading to the house . . .'.

41 The MI5 Lonsdale file for the period 14–26 September 1960 is KV 2/4431.

42 The exact number of women officers in 1960 is not known. There were three women of officer rank in B Division (intelligence) in March 1952: Andrew, *Defence of the Realm*, p. 325. Wright, Spycatcher, pp. 187–8.

43 Ibid., *Spycatcher*, pp. 24–5.

44 It seems the Swiss also sent over a second cipher pad for Wright to examine in early October: note, McBarnet, 4 October 1960: KV 2/4432.

45 See Elwell's Lonsdale 'biography', 18 October 1960: KV2/4383.

46 For Brik see Andrew and Mitrokhin, *Mitrokhin Archive*, pp. 217–23; and Wright, *Spycatcher*, pp. 87–8.

47 See MI5 files on Abel: KV2/3897 and 3898.

48 Contrary to the views of some. See, for example, the anecdote of David Cornwell (John Le Carré) about the head of D1, Courtenay Young, allegedly telling new MI5 recruits around 1959 that they should not waste time looking for Soviet illegals in the UK, because if there were any the Security Service would already know their names. Quoted in Sisman, *John Le Carré*, p. 93, no doubt reflecting Cornwell's scornful view of the Security Service at that time.

49 Herman interview.

50 Evidence of Arthur Bonsall to the Romer Inquiry, 11 May 1961, pp. 3–4: CAB 301/253.

51 See entry on Bonsall in *DNB* by David Omand, updated 15 February 2018. Herman interview. Herman describes Bonsall as 'the cleverest Director GCHQ had' (Bonsall was GCHQ Director 1973–8).

52 Herman interview.

53 Private information, former GCHQ employee.

54 Charles Elwell interview.

55 Declassified MI5 documents for 5 October–21 November 1960 are in KV2/4383 and are the main source for this account of the Portland investigation during this period, except where stated otherwise.

56 Elwell to Smith, 22 February; Smith note, 1 March 1961: KV 2/4476.

57 The MI5 Lonsdale file for 18 to 31 October 1960 is KV 2/4433.

58 KV 2/4434 contains MI5 Lonsdale documents for 31 October–9 November 1960.

59 The account in this book of the bugging of Molody's Albany Street flat and GCHQ monitoring operation relies on the few relevant documents in KV 2/4383 and in KV 2/4434, supplemented by published sources: Andrew, *Defence of the Realm*, p. 487; Wright, *Spycatcher*, pp. 92–4, 133–4; West, *Illegals*, pp. 165–7.

60 Whyte memo, 9 November 1960: KV 2/4434.

61 Wright, *Spycatcher*, p. 134; West, *Illegals*, p. 167.

62 Wright, *Spycatcher*, pp. 134, 137; private information, former GCHQ employee.

63 For details of the MI5/GCHQ operation at Lonsdale's flat see Wright, *Spycatcher*, p. 134.

64 Three pieces of documentary evidence that GCHQ successfully intercepted some of Lonsdale's incoming radio traffic are in the indexes of KV 2/4437 (a note dated 9 December 1960 headed 'Signals received by "Trek"') and KV 2/4439 (a note entitled 'Message from Moscow area to "Trek"' of 1 December 1960); and 'A Note on the Operational Intelligence', p. 3, attached to Hedger note, 8 November 1961 in KV 2/4454. The first two of these notes themselves, however, are still classified. MI5 burgled Lonsdale's flat a second time to examine the lighter on 24 November. Nothing new was found in the hiding place then, but Elwell noted the radio transmission schedule was still there: Elwell note, 25 November 1960, KV 2/4436.

65 KV 2/4434 is the MI5 Lonsdale file for 31 October–9 November 1960.

66 Quote from George Leggett in Observation Briefing Sheet, 2 November 1960, KV2/4383.

67 The intercept itself, dated 27 October 1960, has been withheld from KV2/4383.

68 KV 2/4484 and 4485 are respectively the MI5 files on the Krogers for 29 April 1958–22 November 1960, and 23 November–6 December 1960.

69 According to the renegade MI5 officer, Peter Wright (*Spycatcher*, pp. 132–3), most influential in the decision to move the investigation was his discovery that the Russian embassy had been monitoring the radios of the A4 watchers during the two operations in September to access Lonsdale's safety deposit box. No released MI5 material supports Wright's account. It seems more likely that the principal reason was MI5 management deciding that it was more appropriate for Soviet counter-espionage to take over an investigation focused on a KGB illegal.

70 Andrew, *Defence of the Realm*, pp. 372, 428, 504; Wright, *Spycatcher*, pp. 122–4; Bower, *Perfect English Spy*, pp. 91–2, 136.

71 William Garaint T.P. Colfer, who studied at Oriel College, Oxford, and died in 2014, aged eighty-seven.

72 Wright, *Spycatcher*, pp. 134–5; West, *Illegals*, p. 167.

73 In the words of the MI5 officer, Stewart, in Hugh Whitemore's 1983 play based on events in Ruislip: Whitemore, *Pack of Lies*, p. 5.

74 For Gay Search's reminiscences of events at Ruislip: 'The Russian spies who came in from suburbia', *Sunday Telegraph*, 13 July 2019; 'The spies in a suburban bungalow', BBC News magazine, 11 November 2014 at https://www.bbc.co.uk/news/magazine–29985359

(accessed 27 January 2020). Kroger MI5 files from 4 November 1960 onwards (KV 2/4484), and evidence given at the Portland trial by the MI5 watchers based at 1 Courtfield Gardens (CRIM 1/3604, Parts 3 and 4).

75 Based on an anecdote told by Charles Elwell senior to his son, Charles Elwell. Charles Elwell interview.

76 Charles Elwell interview. Not surprisingly, Elwell did not appear to record his Bill Search conversation in any formal memo. When he conducted a follow-up interview with Ruth Search about a year later on 16 November 1961, Elwell referred to their last meeting 'almost exactly a year ago under slightly different circumstances', suggesting that he may in fact have met Bill and Ruth Search together (or even Ruth Search alone) to persuade them to allow the OP to continue: see para. 2 of Elwell note of meeting with Mrs Search, 20 December 1961: KV 2/4455.

77 It is clear from the MI5 files that in his *Pack of Lies* play, for artistic reasons Whitemore exaggerated the closeness of the Krogers to the Search family, and also the fragility of the Mrs Search character.

78 See Romer Committee Report, pp. 26–7: CAB 301/248; and evidence of Bonsall to the Romer Inquiry, 11 May 1961, pp. 8–15: CAB 301/253.

79 See Romer Committee Report, p. 27: CAB 301/248; evidence of Bonsall to the Romer Inquiry, 11 May 1961, pp. 3–4, 6–7, 18: CAB 301/253.

80 References to events concerning MI5 from 21 November 1960 to 13 January 1961 are based on documents in Houghton's KV2/4384 file, Lonsdale's 4436 to 4440 inclusive, and the Krogers' 4485 and 4486, unless otherwise stated.

81 Evidence of 'Miss K' of 14 March 1961 in CRIM 1/3604, file 3.

82 The MI5 Lonsdale files for the periods 5–17 December and 19 December–28 December 1960 respectively are KV 2/4437 and 4438.

83 Murphy et al., *Battleground Berlin*, pp. 343–6.

84 KV 2/4439 and 4440 are respectively the MI5 Lonsdale files for 28 December 1960 to 5 January 1961, and 6 to 11 January 1961.

85 Murphy et al., *Battleground Berlin*, pp. 343–6, 497; declassified cable, BOB to CIA HQ, 4 January 1961, and memo, 'Activities of 4 and 5 January 1961 and BOB Comments', 15 February 1961 (released by the CIA under FOIA).

86 Hollis evidence to Romer, 4 May 1961, p. 36: CAB 301/253.

87 Kynaston, *Modernity Britain*, pp. 438, 468, 604–5.

88 The account of events on 7 January 1961 is based on MI5

documents in KV2/4384, 4440 and 4487, and trial evidence in CRIM 1/3604, file 3; supplemented by Tietjen, *Spy Ring*, pp. 7–11, 17–20; Bulloch and Miller, *Spy Ring*, pp. 13–17; Lewis, *Spycatcher*, pp. 118–21; Wright, *Spycatcher*, pp. 135–7.

89 Intriguingly, there is no evidence in the declassified MI5 files to corroborate Wright's assertions that Lonsdale entertained a woman overnight in his flat on 6–7 January or that Lonsdale received a coded message from Moscow on the morning of 7 January. The fact that the declassified files do not provide such evidence does not, however, mean that these events did not happen. I have therefore included them here – but with this caveat. Wright, *Spycatcher*, pp. 136–7.

90 See Elwell's file note of 17 January 1961: KV2/4385.

91 Bulloch and Miller, *Spy Ring*, pp. 16–17.

92 The MI5 Kroger file for 28 December 1960 to 10 January 1961 is KV 2/4487.

93 Smith, Ferguson Smith, and Winterbottom evidence in the March 1961 trial: CRIM 1/3604, files 3 and 4; Smith, 'Soviet Spy Ring', pp. 295–9.

94 Kate Dumville interview. Her father, Dr John Sangster, had bought numerous antiquarian books from Kroger.

95 Elwell memo headed 'A Reconstruction', undated but approx. 9 January 1961: KV 2/4487.

96 Bulloch and Miller, *Spy Ring*, pp. 46–7. See Hollis confirmation of how and when the Cohens were first identified by New Scotland Yard: evidence to Romer Inquiry, 11 May 1961, p. 2: CAB 301/253.

Part Two

1 See documents from Morris and Lona Cohen FBI file number 100-406659 released under FOIA (here noted as the 'Cohen FBI file'): https://vault.fbi.gov/Morris and Lona Cohen/Morris and Lona Cohen, accessed 27 January 2020). The Cohen FBI file includes only certain declassified and redacted documents for the period 12 November 1953 to 12 August 1958, and a handful of documents from 9 March 1970 onwards. Carr, *Operation Whisper*, pp. 17 ff.

2 Powers, *Broken*, pp. 221–2, and Powers quoting Hoover before Senate Internal Security Subcommittee, 17 November 1953.

3 Belmont, 'As I recall it', pp. 6, 145–6; memo, Crowl to Tolson, 12 May 1938; memo on Belmont by Tracy, 17 June 1939: FBI Belmont

personnel file released under FOIA (FBI Belmont file), vol. 1, at https://archive.org/details/foia_Belmont_Alan_1 (accessed 27 January 2020). Hoover to Belmont, 11 March 1943, ibid., vol. 2. Conroy to Hoover, 3 May 1945, ibid., vol. 3.

4 Powers, *Broken*, pp. 194–5. Memo, Hendon to Tolson, 20 February 1946; memo, Nease to Hoover, 19 September 1946: FBI Belmont file, vol.3.

5 Andrew, *Secret World*, pp. 672–3.

6 Powers, *Broken*, pp. 202–16.

7 Lamphere and Shachtman, *FBI–KGB War*, pp. 100–120; Belmont, 'As I recall it', p. 145; Powers, *Broken*, pp. 196–7.

8 Belmont, 'As I recall it', pp. 6, 145–6.

9 Lamphere and Shachtman, *FBI–KGB War*, pp. 132–41; Belmont, 'As I recall it', pp. 146–8; memo, Hoover to Tolson, 6 February 1951; memo, 6 October 1951: FBI Belmont file, vol. 3, at https://archive .org/details/foia_Belmont_Alan_3 (accessed 27 January 2020).

10 Lamphere and Shachtman, *FBI–KGB War*, pp. 141–51, 161–2, 175–87, 265. Memo, Hoover to Belmont, 22 June 1953: FBI Belmont file, vol. 4 at https://archive.org/details/foia_Belmont_Alan_4 (accessed 27 January 2020).

11 Memo, Clegg to Tolson, 29 December 1953; memo, Belmont to Boardman, 7 June 1954: Belmont FBI file, vol. 4. Belmont, 'As I recall it', pp. 6–10, 32, 149–53; Carr, *Operation Whisper*, pp. 25–6; Deloach, *Hoover's FBI*, p. 92; Lamphere and Shachtman, *FBI–KGB War*, pp. 12–13, 141. On black agents: https://www.fbi.gov/news /stories/early-african-american-agents (accessed 27 January 2020).

12 *WPA Guide to New York City*, pp. 629 ff.

13 Its external appearance can still be imagined from 169 East 71st Street, a few doors away, where Holly Golightly's apartment was set in the 1961 film *Breakfast at Tiffany's*.

14 FBI report, 26 February 1954: Cohen FBI file; Carr, *Operation Whisper*, pp. 24–5.

15 Memo, Belmont to Boardman, 26 November 1957: https://en.wiki source.org/wiki/%22Operations_of_the_MGB_Residency_at_New _York%2C_1944-45%22 (accessed July 2019); memos, 26 February and 27 December 1954, Cohen FBI file.

16 Memos to Hoover, 12 May 1955 and 19 October 1956, Cohen FBI file.

17 Belmont, 'As I recall it', pp. 138, 154; Deloach, *Hoover's FBI*, pp. 11–12; Lamphere and Shachtman, *FBI–KGB War*, p. 141.

18 This account of the FBI's investigation into the Cohens after 21 June

1957 is based mainly (and unless otherwise indicated in the Notes) on the FBI Cohen files hand-delivered to MI5 on 19 January 1961: KV 2/4489. Letter, Belmont to Tomkins, 14 August 1957, Cohen FBI file.

19 Interview with Sophia Vetrone, 8 August 1957, p. 57: KV 2/4489.

20 Hoover memo, 9 August; memo to Hoover, 16 August 1957: Cohen FBI file. 'Real tigers': description by former FBI agent Robert J. Beatson in Albright and Kunstel, *Bombshell*, p. 247.

21 Memo, Kelly (Special Agent in Charge, New York) to Hoover, 2 August 1957; Hoover memo, 9 August 1957; and memo, 15 August 1957: Cohen FBI file.

22 'Possible Connection of Morris Cohen with the Rosenberg Espionage Operation', various interviews, August 1957, pp. 55–6: KV 2/4489. Memo, 8 January 1957: Cohen FBI file.

23 Albright and Kunstel, *Bombshell*, pp. 18–23, 44, 308–9; Tchikov, *Comment Staline*, p. 289.

24 See FBI Winston interview, 19 November 1957, pp. 72–9; and attempt to locate storage company, p. 62: KV 2/4489.

25 Memos, 25 July, 1 and 13 August 1957 Cohen FBI file.

26 Albright and Kunstel, *Bombshell*, pp. 26–9. The FBI found out subsequently that Cohen was issued with a third passport in yet another false name by the US consul general in Barcelona so he could return to the USA without his part in the Spanish Civil War becoming known: Romer Report, p. 47: CAB 301/348; memos, 14 and 21 February 1958: Cohen FBI file. Svirin and haemorrhoids: Lonsdale Report, p. 54: KV 2/4466.

27 Interview with Frank De Scipio, 19 August 1957, pp. 52–3: KV 2/4489.

28 Interview with Anthony Vetrone, 2 December 1957, p. 83: KV 2/4489.

29 Interview with Sophia Vetrone, 18 November 1957, pp. 69–70: KV 2/4489.

30 Interview with Ella Cohen, 5 December 1957, pp. 11–12: KV 2/4489. Memos, 11, 17 and 22 February 1958: Cohen FBI file.

31 Memo, 21 February 1958: Cohen FBI file. Cimperman (US legal attaché, London) to Young, 29 April 1958; Requests for Information Regarding Aliens, 14 October and 10 November 1958: KV 2/4484. Hollis evidence to Romer, 25 May 1961, pp. 2, 19: CAB 301/253.

32 Beatson quoted in Albright and Kunstel, *Bombshell*, p. 247

33 Beatson, ibid. Memos, 21 February, 11 and 17 August 1958: Cohen FBI file.

34 See documents dated between 14 October 1958 and 2 October 1959: KV 2/4484.

35 Interview with Nathan Kaplan, 13 November 1957, pp. 98–100: KV 2/4489.

36 The FBI reports were received at Leconfield House on 19 January 1961: KV 2/4489.

Part Three

1 The account of events in this section from 9 to 13 January 1961 is based on KV2/4384, and from 16 January 1961 to 10 February 1961 on KV2/4385, unless otherwise stated.

2 Undated manuscript memoir headed 'The Houghton and Gee Affair' by an anonymous engineer who joined the Type 2001 sonar project in 1957: Portland Museum collection.

3 Report of a visit to AUWE Portland on 11–12 January 1961, by Commander A.G. Skipwith of MI5, 20 January 1961: CAB 301/254.

4 Smith, 'Soviet Spy Ring', pp. 299–301; Smith evidence at trial, 14 March 1961: CRIM 1/3604, Part 3, pp. 53 ff.

5 Elwell note, 17 January 1961: KV2/4385.

6 Winterborn note, 'Search for Transmitter at Ruislip', 13 January 1961: KV 2/4441; Ferguson Smith evidence at trial, 15 March 1961: Part 4, CRIM 1/3604; Smith, 'Soviet Spy Ring', pp. 303–4; Bulloch and Miller, Spy Ring, p. 39; Wright, Spycatcher, p. 138.

7 Bonsall evidence to Romer, 11 May 1961, pp. 3–7: CAB 301/253.

8 'Spies, lies and a play that slips up', Evening Standard, p. 8, 2 December 1983: Morris Cohen papers.

9 Smith evidence at trial, 13 March, Part 3, pp. 58–65, 81: CRIM 1/3604.

10 KV 2/4441 and 4442 are the MI5 Lonsdale files covering the periods 12–19 January, and 20 January–9 February 1961 respectively. Wright, Spycatcher, p. 138.

11 Hollis note, 17 January 1961: KV 2/4441.

12 Hoover to Hollis, 28 March 1961: KV 2/4447.

13 Interview Manningham-Buller; Sampson, Anatomy of Britain, p. 151.

14 Sampson, Anatomy of Britain (as a result of the Profumo affair the UK intelligence services merited several pages in Sampson's revised edition, Anatomy of Britain Today, published in 1964); Aldrich and Cormac, Black Door, p. 234.

15 Bulloch and Miller, Spy Ring, p. 154

16 Ibid., pp. 154 ff; Hennessy, *Prime Minister*, p. 264; Aldrich and Cormac, *Black Door*, pp. 206–7; Catterall, *Macmillan Diaries*, diary entry 7 May 1960, pp. 291–2.

17 Minutes, Macmillan to Hume, 1 and 31 August 1960; note of JIC meeting, 15 September 1960: PREM 1/4721. A memo from James Robertson, Secretary to the Cabinet, 25 October 1960, confirms that 'following the U-2 and RB-47 incidents, the prime minister has had to devote a good deal of time to intelligence matters': PREM 11/3101. Letter, Acland to de Zulueta, 15 August 1960 (setting out conclusions of Working Party chaired by MI5 on the discovery of microphones in the Moscow embassy in October 1959, and noted by Macmillan): PREM 11/3104. Manningham-Buller consents: CRIM 1/3604, Part 1.

18 Gaddis, *The Cold War*, pp. 73–5.

19 Memo, Admiralty Permanent Secretary to Commander-in-Chief Far East, 9 March 1961: ADM 1/27960.

20 Stone to Elwell, 19 January 1961; and Stone to Hill (extract), 31 January 1961: KV 2/4490.

21 Gouzenko: RCMP to MI5, 13 February 1961: KV 2/4444; Petrovs: ASIO to MI5, 1 March 1961: KV 2/4446.

22 'Liaison with Allied Security Services', 5 April 1961: KV 2/4446.

23 Smith note, 31 January 1961; Elwell to Special Branch, 3 February 1961: KV 2/4442. Memo, RCMP to Elwell, and Bates to Elwell, 17 February 1961: KV 2/4444. The MI5 Lonsdale files for the periods 10–23 February, 24 February–6 March and 9 March–6 April 1961 are respectively KV 2/4444, 4445 and 4446.

24 Smith 'biography' of Krogers in CRIM 1/3604, Part 1; Manningham-Buller evidence to Romer Inquiry, 11 April 1961, pp. 14–16: CAB 301/252; copy of Kroger passport application papers, Romer: CAB 301/254; note by Elwell, 30 January 1961: KV 2/4490.

25 Bulloch and Miller, *Spy Ring*, pp. 154 ff.

26 See Bowers's witness statement in CRIM 1/3604, Part 2. Details of the March 1961 trial are based on documents in CRIM 1/3604, including the verbatim transcript of proceedings, unless otherwise indicated.

27 Hollis evidence to Romer, 25 May 1961, p. 6: CAB 301/253.

28 MI6 to Elwell, 16 March, and Martin note, 28 March 1961: KV 2/4477. Draft Smith statement (undated): CRIM 1/3604, Part 1.

29 Snelling, *Rare Books*, pp. 206–9.

30 Furnival Jones to Hollis, and attached Martin note, 'Summary of Security Information', 6 March 1961: KV 2/4476.

31 Elwell telegram to SLO Malaya, 10 March; Security Intelligence Far East telegram to MI5 head office, 14 March 1961: KV 2/4477.

32 Memo, Carrington to Macmillan, 9 March 1961: ADM 1/27960; Ormsby-Gore to Macmillan (extracts), 1 March 1961: CAB 301/249.

33 Various memos, 9 and 10 March 1961: ADM 1/27960.

34 This account of the trial is based on: the verbatim transcripts found in CRIM 1/3604, Parts 3 to 6; Bulloch and Miller, *Spy Ring*, pp. 187–216; Tietjen, *Soviet Spy Ring*, pp. 82–137; except where otherwise made clear.

35 Draft order for court to go into *in camera* session, undated: CRIM 1/3604, Part 1.

36 CRIM 1/3604, Part 4 and following.

37 Note, Synnott to Carrington, 17 March 1961; notes, Synnott to Carrington, 19 March 1961; note of meeting, 20 March 1961: ADM 116/6295.

38 Tietjen, *Soviet Spy Ring*, p. 110.

39 Lonsdale Preliminary Report, March 1961, p. 15: KV 2/4466.

40 Morris Cohen later confirmed this explanation of Lonsdale's statement at the trial: Albright and Kunstel, *Bombshell*, p. 252.

41 Bulloch and Miller, *Spy Ring*, p. 211.

42 Manningham-Buller evidence to Romer, 11 April 1961, pp. 9–10: CAB 301/252.

43 Davies to Home Office, 20 March 1961; note by Synnott, 21 March 1961; Butler to Manningham-Buller, 21 March 1961: ADM 1/27960. Furnival Jones to Admiralty, 20 March 1961: ADM 116/6295.

44 Memo, Bishop to Brook (with MS addition), 22 March 1961: CAB 21/6012.

45 Macmillan diary entry, 24 March 1961: MSMacmillan dep. d.41.

Part Four

1 Hansard, HC Deb 23 March 1961, cols. 584–94.

2 Synnott to Carrington, 13 March 1961: ADM 1/27960.

3 MI5's 'Security in the United Kingdom and the USA', 20 March 1961, para. 5: CAB 301/249. John to Burke, 15 March 1961: ADM 1/27960.

4 Preston, 'Submarine Design'; Wright, 'Aircraft Carriers', pp. 149–50, 170–75.

5 Parry, 'History', in the section headed 'The Nuclear Revolution, Sonar 2001'; Parry interview. An Introduction to the Admiralty Underwater Weapons Establishment Portland, 1963, pp. 1–2, 8–9:

ADM 302/147; Scientific and Technical Progress Report 1961,
AUWE Portland, pp. 1–15, 19–22: ADM 302/141; Weapon Control
System for Nuclear Submarines, UDE pamphlet no.653, October
1959, pp. 1–2: ADM 259/532; Introduction to Sonar Type 2001,
1963: ADM 302/327.

6 Hennessy, *Secret State*, p. 2.

7 Ring, *We Come Unseen*, pp. 68–74; West and Tsarev, *Crown Jewels*,
p. 258; Polmar and Whitman, *Hunters and Killers*, pp. 75–86;
Polmar and Moore, *Cold War Submarines*, p. 83.

8 Stephenson to Hollis, 6 March 1961, CAB 301/249; MI5's 'Security
in the United Kingdom and the USA', 20 March 1961, para. 6: CAB
301/249.

9 Macmillan diary entry, 26 March 1961: MSMacmillan dep. d.41.

10 Macmillan to Carrington, 23 March 1961; Brook to Butler, 24
March 1961: CAB 21/6012. Note to AUWE Portland, 24 March
1961: ADM 116/6295. Thorpe, *Supermac*, pp. 491–96.

11 Kennedy to Martin, 28 March; Martin to Macaffee, 29 March; Hill
note, 7 April 1961: KV 2/4477.

12 Trend to Brook, 28 March 1961: CAB 301/249; note of 10 April 1961
meeting: CAB 301/249.

13 Here termed the Lonsdale Report, March 1961: KV 2/4466.

14 Appendix C, Lonsdale Report; Appendix J of Romer Report, p. 46,
CAB 301/248.

15 The full text of the Lonsdale letters is in Appendix C of the Lonsdale Report, March 1961: KV 2/4466.

16 Ibid., Appendices A and B, KV /4466. The Lonsdale Report did not
include any reference to GCHQ's interception of Lonsdale's wireless
traffic to and from Moscow before the ring were arrested. Most
details of this remain classified, but it is clear that, for example,
GCHQ was aware of a transmission made to Lonsdale on 7 October
1960 in accordance with his wireless schedule, Elwell letter to
GCHQ, 7 April 1961: KV 2/4447. On deciphering the 15 January
1961 message see Hollis to Kent and attachment, 11 January 1963:
KV 2/4458.

17 Admiralty evidence to Romer, 19 April 1961, pp. 24, 35; Austen
evidence to Romer, 21 April 1961, pp. 1, 16; evidence of Watford,
Yarram and Kemley to Romer Inquiry, 28 April 1961: CAB 301/252;
and of Crewe-Read, 2 May 1961; Hollis, 4 May 1961, pp. 1–2, 22;
and of Denning, 9 May 1961, pp. 1–2: CAB 301/253.

18 Johnson evidence to Romer, 16 April 1961, pp. 4–5: CAB 301/253.

19 Bonsall evidence to Romer, 11 May 1961, pp. 1–7, 17: CAB 301/253.

20 Hollis evidence to Romer, 18 April 1961, pp. 1, 3–5, 11–12, 22: CAB 301/252.

21 Hollis evidence to Romer, 4 May 1951, pp. 11, 15–18, 33; and 25 May, pp. 21, 25: CAB 301/253.

22 Hollis evidence to Romer, 4 May 1961, p. 29; 11 May 1951, pp. 2–3, 5; and 25 May 1961, pp. 12–14, 16, 21–2: CAB 301/253.

23 Bower, *Perfect English Spy*, p. 267; Horne, *Macmillan*, pp. 456–7; Hennessy, *Winds of Change*, p. 245.

24 Andrew, *Defence of the Realm*, pp. 484, 489; Carr, *Operation Whisper*, p. 273.

25 Cabinet minutes, 4 and 11 May 1951; Macmillan to Romer and response, 15 May 1961; Hunt to Brook, 30 May 1961: CAB 301/250.

26 Kynaston, *Modernity Britain*, pp. 565, 571, 576, 584; Davenport-Hines, *An English Affair*, pp. 115, 122.

27 Macmillan interview.

28 Nicholson, *How Was It For You?*, p. 2.

29 Report of the [Romer] Committee of Inquiry into Breaches of Security, 1961, paras. 50, 52, 90: CAB 301/248.

30 Ibid, paras. 71–9, 87–90.

31 Carrington, *Reflect on Things Past*, p. 171; Hansard, HC Deb 13 June 1961, cols. 211–17.

32 Robertson to Hollis, 16 June 1961; Mitchell to Hollis, 19 June 1961: CAB 21/6012.

33 Hansard, HC Deb 22 June 1961, cols. 1683–1688; HL Deb 5 July 1961, cols. 1418–1420.

34 See ADM 1/28160, 1/28308 and 1/30088.

35 FBI Reports, 12 June and 24 November 1961: Robert F. Kennedy Attorney-General Papers, National Security Classified File, 13-01-13, NATO Section 1, 20 January 1961–19 October 1962 (declassified 2019). Hewitt traces the history of the abuses of Canadian passports in his 2008 article in *Intelligence and National Security*. On the Lonsdale passport see Hewitt, pp. 393–4. Memo, 'Canadian Passports', Sweeny to DSI, 5 October 1961, declassified May 2019, FOI A-2018-00826/PG (the vast majority of the RCMP file on Gordon Lonsdale remains classified).

36 Report of the Committee on Security Procedures in the Public Service, Part 2, para. 25, p. 6: CAB 301/258; Maude to Trend, 17 August 1961: CAB 301/260; Andrew, *Defence of the Realm*, p. 955, fn. 50.

Part Five

1 Hollis evidence to Romer, 11 May 1961, pp. 7, 9: CAB 301/253.

2 Report in *The Times*, 8 May 1961: KV 2/4448. The MI5 Lonsdale files for 1961 are KV 2/4447 (6 April–2 May), 4448 (2 May–13 June), 4449 (14 June–7 July), 4450 (7–21 July), 4451 (21 July–15 August), 4452 (15 August–14 September), 4453 (15 September–25 October), 4454 (26 October–16 November), and 4455 (17 November 1961–12 January 1962). All references to MI5 documents in this Part concerning 1961 are to papers in these Lonsdale files unless otherwise stated.

3 MI6 to Martin, 24 May 1961: KV 2/4448.

4 Blake, *No Other Choice*, pp. 213–14; Snelling, *Rare Books*, pp. 222, 228; Hermiston, *Traitor*, pp. 263–4. Maude to Trend, 8 June 1964: CAB 301/403.

5 Elwell notes, 28 April; and 2, 8 and 29 May 1960.

6 Elwell's 'Considerations on Possible Negotiations', 3 May; Hill and Hollis notes, 5 and 25 May 1961.

7 Elwell note of interview, 1 June; and extra notes, 2 June 1961.

8 Elwell notes, 15 June 1961.

9 Elwell note of Lonsdale interviews on 21 and 22 June 1961.

10 Lonsdale to Baker, 21 April 1961: KV 2/4447. Elwell notes, 10 and 29 May; FBI letter, 16 May; MI6 to Elwell, 16 May: KV 2/4448.

11 Molody's first name is spelt Conon in FBI documents but in this book, for consistency with most Western and Russian sources, is Konon.

12 Elwell to Minnich, 30 June; Bates to Elwell, 3 July 1961.

13 Bates to Elwell, 11 July; FBI Airtel, 7 July; Elwell to MI6, 14 July; Elwell note, 17 July 1961.

14 Cunningham to Hollis, 19 July; Elwell note, 31 July 1961.

15 Elwell notes, 10, 11, 16 August 1961; and transcripts headed 'Interrogation of Lonsdale and Mr and Mrs Kroger' and 'Conversation between Lonsdale and Mr Elton' (serial 1094A).

16 Elwell notes, 17, 20, 24 July; Stone to Martin, 31 July.

17 FBI Tatiana Piankova interview report, 22 September 1961.

18 Elwell note, 29 September; Elwell 'Note on Visit by D1/CJLE to Paris', 9 October; Bates to Elwell and interview with Piankova attached, 10 October 1961.

19 Elwell notes, 16 and 20 November 1961.

20 Elwell, 'Procedure for Further Negotiations', 16 October 1961.

21 Martin to Bates, 27 October; Elwell note, 9 November 1961.

22 *Daily Telegraph*, 25 November 1961, and other newspaper cuttings in KV 2/4455.

23 Elwell note, 1 December 1961.

24 Hedger to Winterborn and attachments, 8 November 1961; Elwell note, 8 December 1961. The Cabinet and prime ministerial papers documenting approval for the Lonsdale deal were still classified when this book went to press.

25 Elwell note, 8 December 1961.

26 Elwell note of Lonsdale interview on 9 January 1962.

27 The MI5 Lonsdale files for the period 15 January–22 March, 27 March–18 December and 31 December 1962–2 April 1963 are respectively KV 2/4456, 4457 and 4458. All references to MI5 documents in this Part concerning 1962 and up to 2 April 1963 are to papers in these Lonsdale files unless otherwise stated.

28 Bates to Martin, 26 April 1962 and 12 April 1962 FBI attachment, 'Gordon Arnold Lonsdale'.

29 Martin and Hill notes, 1, 16 February, and Small note, 9 March 1962.

30 MSS Macmillan dep. d.48–50: diary entry for 11 January 1963.

31 Small to Minnich, 31 January; Buczkowksi to Lonsdale, 9 February 1961. MI5 Lonsdale files for 3 April–20 August 1963, 23 August–18 December 1963, 19 December 1963–7 April 1964, 7 April–10 July 1964 and 14 July 1964–12 March 1965 are respectively KV 2/4459, 4460, 4461, 4462 and 4463. All references to MI5 documents in this Part concerning this period are to papers in these Lonsdale files unless otherwise stated.

32 Furnival Jones note 1443 to Hollis, 21 February.

33 Documents detailing Cabinet-level involvement with the spy exchange for the period 9 May 1964–2 October 1964 are in CAB 301/403 (all references confirming Wynne was spying for MI6 are redacted out). Lonsdale, *Spy*, pp. 195–9.

34 Note, FCO London to Moscow embassy, 6 April 1964; FCO Statement, 22 April 1964: FO 953/2264.

35 Clayton (who accompanied Lonsdale to RAF Northolt on 21 April 1964) wrote a vivid account of the day: note, 7 May 1964.

36 Note headed 'Wynne', Hadow to Caccia, 22 April 1964; covering note, Hill to Hadow, 24 April 1964: FO 953/2264. Security Service leak report by Hill, 23 April 1961, and related documents: CAB 301/403.

37 Lonsdale, *Spy*, pp. 202–13.

38 Lonsdale to Helen Kroger, 30 April; Peter Kroger to Lonsdale,

12 May; 'Maria Petka' to Helen Kroger, 25 May 1964: KV 2/4462. MI5 Lonsdale files for 7 April–10 July 1964, 14 July 1964–12 March 1965, 18 March 1965–18 February 1966 and 21 February 1966–1 April 1967 are respectively KV 2/4462, 4463, 4464 and 4465. All references to MI5 documents or prison correspondence in this Part concerning this period are henceforth to papers in these Lonsdale files unless otherwise stated.

39 Andrew, *Defence of the Realm*, p. 488; interview Charles Elwell.

40 Tchikov, *Comment Staline*, p. 321.

41 Dumville interview.

42 Snelling, *Rare Books*, pp. 215–24, 239–42.

43 CAB 301/254.

44 See Hollis/Ruck evidence, 11 May 1961, p. 5: CAB 301/253.

45 The MI5 files on Desmond Costello are KV2/4328 (3 October 1934–27 January 1951), 4329 (29 January 1951–13 March 1959), and 4330 (6 April 1959–20 May 1963). They are the source for the sections of this book on the Security Service investigation into the Costellos, except where stated.

46 Zohrab, who was of Armenian extraction, died in 2008. Presumably he was investigated further by the NZSS and cleared of any suspicion because he went on to have a distinguished career as a public servant, culminating in his appointment as ambassador of New Zealand to Germany, 1969–74. See: https://en.wikipedia.org/wiki/Doug_Zohrab (accessed 12 April 2019).

47 Lamphere and Shachtman, *FBI–KGB*, p. 292; Andrew and Mitrokhin, *Mitrokhin Archive*, pp. 534, 864 (fn. 73). The KGB Paris residency sent an encrypted telegram to the Centre in November 1953, seven months before Morris Cohen applied to the Paris consulate for New Zealand passports. Responding to enquiries from Moscow, the Paris residency confirmed that it had 'established' how the 'Volunteers' (the Cohens' former KGB code-name) could safely apply for their passports by post. See Tchikov, *Comment Staline*, p. 234.

48 The MI5 files on the Krogers which were released when this book went to press end on 6 February 1961. FBI estimate that their (still) classified Cohen files for the period 1959–1970 contain around 19,000 documents: FBI letter to author, 11 March 2019.

49 Kroger and Lonsdale, *Pisma, passim*.

50 Letters to Myers from: Helen Kroger, 3 January 1968; Peter Kroger, 3 January, 6 February, 26 February, 20 August 1968: Morris Cohen papers.

51 Quoted in Platt, 'Soviet Imprisonment', p. 197. This account of the

British government's approach to the Brooke/Krogers exchange is based on Platt, 'Soviet Imprisonment', except where indicated.

52 Lonsdale to Helen Kroger, 3 August 1965: KV 2/4464. MI5 Lonsdale files for 18 March 1965–18 February 1966, and 21 February 1966–1 April 1967 are KV 2/4464 and 4465 respectively. All references to MI5 documents or prison correspondence in this Part concerning this period are to papers in these Lonsdale files unless otherwise stated.

53 Lonsdale to Peter Kroger, 3 August 1965 and 16 March 1966. Tchikov, *Comment Staline*, p. 282.

54 Platt, 'Soviet Imprisonment', pp. 197–202.

55 Ibid., pp. 203–6.

56 Letter, Peter Kroger to Myers, 29 September 1969, Morris Cohen papers; 'The Times Diary': 'Kroger's Farewell Supper', *The Times*, 15 September 1969.

57 Platt, 'Soviet Imprisonment', pp. 206–7.

58 Gee to Houghton, 21 March 1965: KV 2/4474.

59 Houghton, *Operation Portland*, pp. 124–39.

60 Elwell note, 'History of Houghton's Espionage', 1 May; and file notes, 23 and 25 May 1961: KV 2/4478. Other sources confirming Houghton's 1951 recruitment: Elwell evidence, extract from Vassall Tribunal transcript, p. 26: KV 2/4481.

61 Elwell file note, 25 May 1961: KV 2/4478.

62 Jupp to Elwell and attachment, 14 August; Elwell notes, 15 December 1961: KV 2/4479.

63 Hollis notes, 8 and 20 November; Elwell to D1, 19 November 1962: KV 2/4480.

64 Hill notes, 8 and 9 January 1963: KV 2/4480. Extracts from transcript of Vassall Tribunal, and report (see para. 269): KV 2/4481.

65 Young to MacDonald, 11 March; Patrick note, 28 March 1963: KV 2/4481. Patrick note, 29 April 1964: KV 2/4482.

66 Top-secret 'Robe' transcript, 18 September; interview with Elder, 5 May 1964: KV 2/4482. Note by Small, 7 January 1966: KV 2/4474.

67 Houghton, *Operation Portland*, pp. 140–53.

68 Note, Callaghan to Wilson, 1 August 1969: PREM 13/2917.

Part Six

1 Dolgopolov, *Glavny*, p. 220; *Legendarnye*, pp. 333–4; Kolosov and Molody, *Mertvy sezon*, pp. 202–6; Lonsdale, *Spy*, p. 213; Arthey, *Abel*, p. 192; private information from a source in Russia.

2 The statue was toppled on 22 August 1991.

3 Cherkashin, *Spy Handler*, p. 55; Molodaya, *Istoriya*, p. 5 (Natalia Molodaya was Molody's aunt).

4 Undated diary entry, Serov, *Zapiski*, pp. 565–6. Doubts have been voiced about the authenticity of some Serov diary entries concerning episodes in the Second World War and Raoul Wallenberg, but not the ones referred to here.

5 Lyubimov interview.

6 Kolosov and Molody, *Mertvy sezon*, pp. 11–12, 208–210; interviews, Lyubimov and NATO country source; Lonsdale, *Spy*, p. 199.

7 Interviews, Lyubimov and NATO country source.

8 Lonsdale, *Spy*, pp. 214–15; Antonov, *Konon Molody*, pp. 159–60; Eleanor and Kim Philby quotes from footnote on p. 281 of Tchikov, *Comment Staline*.

9 Lonsdale, *Spy*, pp. 10–11, 47–8, 51, 96, 214–15; Kolosov and Molody, *Mertvy sezon*, p. 210; Elwell, 'A Russian Intelligence Officer Exposed'.

10 Lyubimov interview; Andrew and Gordievsky, *KGB*, pp. 365, 515; Antonov, *Konon Molody*, p. 331.

11 Lyubimov interview; Andrew and Gordievsky, *KGB*, p. 399.

12 Andrew and Mitrokhin, *Mitrokhin Archive*, p. 537; Kolosov and Molody, *Mertvy sezon*, pp. 6–8.

13 Blake, *No Other Choice*, pp. 264–5; Dolgopolov, *Legendarnye*, pp. 333–4; Kolosov and Molody, *Mertvy sezon*, pp. 12–15.

14 Andrew and Mitrokhin, *Mitrokhin Archive*, p. 537; Kolosov and Molody, *Mertvy sezon*, pp. 214–16; Dolgopolov, *Legendarnye*, p. 336.

15 Andrew and Gordievsky, *KGB*, p. 367; NATO country source, interview; Andrew and Mitrokhin, *Mitrokhin Archive*, pp. 537–8.

16 Rimington, *Open Secret*, pp. 232–40; Rimington, talk at The National Archives, Kew, 5 April 2019.

17 Private information to the author.

18 Yakunin, *Treacherous Path*, pp. 7, 70.

19 Aldrich, *Hidden Hand*, p. 6.

20 As at May 2020 when this book went to press.

21 See e.g. Russian TV channel Zvezda documentary on Molody (undated): https://www.youtube.com/watch?v=TqAL2iRGDU4; and Chikov, *Sekretniye*, p. 83 fn.

22 Primakov, *Ocherki istorii*, p. 180; Antonov, *Konon Molody*, pp. 27–38; Dolgopolov, *Legendarnye*, p. 326. The Russian sources confirm Molody was not recruited by the KGB while he was serving in the

Red Army, and that he did not encounter Willie Fisher (Rudolf Abel) during this period.

23 Kolosov and Molody, *Mertvy sezon*, pp. 25–30; Carr, *Operation Whisper*, p. 40.

24 Antonov, *Vokopakh*, p. 115; Pavlov, *Operatsiya*, pp. 110, 143–4; Pavlov, *Tragedii*, pp. 282, 285; Primakov, *Ocherki istorii*, p. 180; information from Mikhail Lyubimov.

25 Macintyre, *The Spy and the Traitor*, pp. 18–19.

26 Dolgopolov, *Legendarnye*, p. 328; Antonov, *Konon Molody*, pp. 43–4, 47, 51, 207–8; Agranovsky, *Razvedchik*, p. 6; Tchikov, *Comment Staline*, p. 238.

27 Kolosov and Molody, *Mertvy sezon*, pp. 63–6.

28 Cf. Antonov, *Konon Molody*, pp. 43–4; Primakov, *Ocherki istorii*, pp. 181–3; and Evseev et al., *Moya professiya* (published in 1990 with Molody's wife as one of the authors, this is allegedly based on various interviews with Molody and articles about him prepared in the last year or two before his death in 1970 for a KGB-inspired biography that was never published: Andrew and Mitrokhin, *Mitrokhin Archive*, p. 537).

29 Andrew and Gordievsky, *KGB*, pp. 283–4; Antonov, *Konon Molody*, pp. 66–7; Primakov, *Ocherki istorii*, pp. 181–3.

30 Kolosov and Molody, *Mertvy sezon*, pp. 63–6; Pavlov, *Tragedii*, p. 283; Andrew and Mitrokhin, *Mitrokhin Archive*, p. 532.

31 Agranovsky, *Razvedchik*, pp. 55–7; Andrew and Mitrokhin, *Mitrokhin Archive*, pp. 365, 532; Hewitt, 'Canadian Passport Security', pp. 388–9.

32 Kolosov and Molody, *Mertvy sezon*, pp. 63–70, 79–81, 87–8; Andrew and Gordievsky, *KGB*, p. 366; Pavlov, *Operatsiya*, p. 142.

33 Tchikov, *Comment Staline*, pp. 237–8. There is corroborative evidence of these Molody/Cohens meetings in Paris in the Gordon Lonsdale Canadian passport and the Krogers' New Zealand passports (CRIM 1/3604); and Romer Report, p. 46: CAB 301/248.

34 Tchikov, *Comment Staline*, pp. 283–4.

35 Andrew and Mitrokhin, *Mitrokhin Archive*, p. 536; Tchikov, *Comment Staline*, pp. 284–5; Barsky, *Deep Undercover*, p. 132.

36 Undated UK newspaper cutting headed 'Letters from a master spy', summer 1984: Morris Cohen papers; Tchikov, *Comment Staline*, pp. 284–5.

37 Andrew and Mitrokhin, *Mitrokhin Archive*, pp. 536–7.

38 Tchikov, *Comment Staline*, p. 285; private information from source in Russia.

39 Barsky, *Deep Undercover*, pp. 131–5.

40 Peter Kroger to Myers, 4 November 1982: Morris Cohen papers.

41 Tchikov, *Comment Staline*, p. 285.

42 Tchikov, *Comment Staline* is a translation into French of a manuscript in English. This manuscript was in turn a translation into English by the American Gary Kern from an original in Russian prepared by Tchikov, from which Kern removed the most egregious examples of pro-KGB propaganda. There was also a German edition, *Perseus: Spionage in Los Alamos*, in 1996. See Gary Kern February 2006 blog (accessed June 2019) at: https://lists.h-net.org /cgi-bin/logbrowse.pl?trx=vx&list=H-HO-AC&month=0602&week=c&msg=Lb%2BREHU-oud/%2BSFUeQbPTMA&user=&pw=; information from Kern.

43 Albright and Kunstel, *Bombshell*, pp. 273–4; KGB 1989 Kroger interviews in Tchikov, *Comment Staline*, pp. 287–325.

44 Albright and Kunstel, *Bombshell*, pp. 270, 273–4; private information from a source in Russia; Andrew and Gordievsky, *Instructions*, pp. 108–9; Carr, *Operation Whisper*, p. 288.

45 Private information from a source in Russia; Carr, *Operation Whisper*, p. 289; Prokopenko, *Yaderny*, pp. 43–4.

46 Tchikov, *Comment Staline*, pp. 329–30; Dolgopolov, *Legendarnye*, pp. 313, 323; private information from a source in Russia.

47 Albright and Kunstel, *Bombshell*, pp. 29–31; Carr, *Operation Whisper*, pp. 64–89; Tchikov, *Comment Staline*, pp. 108, 289–90. Vladimir [T]Chikov, a former KGB colonel, is the only published author who has had access to the KGB files on the Cohens: Chikov interview.

48 Albright and Kunstel, *Bombshell*, pp. 32–4, 45–6; Carr, *Operation Whisper*, pp. 102–10; Tchikov, *Comment Staline*, p. 105; Semyonov report to Fitin, 29 November 1944: Vassiliev, White Notebook no. 1, pp. 109, 112; and report headed 'Veterans of International Brigades who took part in the civil war in Spain', undated but early 1940: Vassiliev, Black Notebook, p. 168.

49 Tchikov, *Comment Staline*, pp. 105–6, 293, 295; Albright and Kunstel, *Bombshell*, pp. 46–7.

50 Tchikov, *Comment Staline*, pp. 125–7; Albright and Kunstel, *Bombshell*, pp. 47–8; Weinstein and Vassiliev, *Haunted Wood*, p. 207.

51 Tchikov, *Comment Staline*, pp. 127–30; Albright and Kunstel, *Bombshell*, pp. 48–9, 310, fn. 48–9.

52 Albright and Kunstel, *Bombshell*, pp. 49–50; Carr, *Operation Whisper*, pp. 126–7.

53 Letter NY KGB station to Centre, 17 February 1945: Vassiliev, Black Notebook, p. 120; Semyonov report to Fitin, 29 November 1944: Vassiliev, White Notebook no. 1, pp. 110, 112; Haynes et al., *Spies*, p. 350.

54 Albright and Kunstel, *Bombshell*, p. 50; Carr, *Operation Whisper*, pp. 129–33, 166–72.

55 Tchikov, *Comment Staline*, pp. 294–5, 298; Albright and Kunstel, *Bombshell*, pp. 71–4.

56 Note on 'Volunteer's wife' i.e. Lona Cohen, in Semyonov report to Fitin, 29 November 1944: Vassiliev, White Notebook no. 1, p. 113; entry on 'Link', Vassiliev Notebooks Concordance, p. 246; Carr, *Operation Whisper*, pp. 135–7; Albright and Kunstel, *Bombshell*, p. 71; Haynes et al., *Spies*, pp. 398–405.

57 Note on 'Volunteers', i.e. Morris Cohen and his wife, in Semyonov report to Fitin, 29 November 1944: Vassiliev, White Notebook no.1, pp. 112–13; Haynes et al., *Spies*, pp. 421–2.

58 Albright and Kunstel, *Bombshell*, pp. 94–9; letter KGB NY residency to Centre, 19 March 1945: Vassiliev, Black Notebook, p. 136. On Seborer: see 'Project SOLO and the Seborers', *Studies in Intelligence*, vol. 63, no. 3 (September 2019) at https://www.cia .gov/library/center-for-the-study-of-intelligence/csi-publications /csi-studies/studies/vol-63-no-3/pdfs/Fourth-Soviet-Spy-Los Alamos.pdf

59 Note Fitin to Kvasnikov, 23 February 1945: Vassiliev, Black Notebook, p. 133; and ibid., p. 135, list of Yatskov agents. Carr, *Operation Whisper*, pp. 155–6; Albright and Kunstel, *Bombshell*, pp. 133–4; Russian TV documentary, Zvezda, *Mr and Mrs Cohen: special agents who saved the world*, broadcast 3 November 2016; Andrew and Mitrokhin, *Mitrokhin Archive*, p. 174. Lona Cohen as Fuchs contact: KGB NY residency to Centre, 23 February and 2 March 1945: Vassiliev, Yellow Notebook, no. 1, pp. 18–19.

60 Weinstein and Vassiliev, *Haunted Wood*, pp. 206–11; letter KGB NY residency to Centre, 26 June 1945: Vassiliev, Black Notebook, p. 136; Haynes et al., *Spies*, pp. 115–16; KGB NY residency to Centre, 30 June 1945: Vassiliev, Yellow Notebook no. 1, pp. 26–7.

61 Albright and Kunstel, *Bombshell*, pp. 145, 336 (fn. p. 148 notes seven differing accounts by KGB veterans of Lona Cohen's trip to meet Ted Hall in July 1945). For various versions see: Tchikov, *Comment Staline*, pp. 158–71; Albright and Kunstel, *Bombshell*, pp. 148–53; Primakov, *Ocherki*, pp. 194–5; Andrew and Mitrokhin, *Mitrokhin*

Archive, p. 174; Haynes et al., *Spies*, pp. 115–17. The author's account is an amalgam, largely based on the first three of these sources.

62 Report by Kvasnikov to Merkulov (close associate of Beria), 12 September 1945: Vassiliev, White Notebook no.1, p. 118. This document confirms the rendezvous with Hall took place on a Saturday, not a Sunday.

63 Pavlov, *Tragedii*, p. 292.

64 Haynes et al., *Spies*, pp. 109, 117; Albright and Kunstel, *Bombshell*, pp. 180, 182–3; Tchikov, *Comment Staline*, pp. 192–203; Centre to Sokolov, 19 April 1948: Vassiliev, Black Notebook, p. 127.

65 Albright and Kunstel, *Bombshell*, pp. 168, 184–92; Tchikov, *Comment Staline*, pp. 204–5, 213–14, 360; Haynes et al., *Spies*, p. 124; Andrew and Mitrokhin, *Mitrokhin Archive*, p. 194. Centre to Sokolov, 27 April 1948: Vassiliev, Black Notebook, p. 127; Centre to Uglov (NY), 8 June 1948, ibid., p. 128; Centre to BOB, 5 October 1948, ibid., p. 130.

66 Andrew and Mitrokhin, *Mitrokhin Archive*, p. 194; Albright and Kunstel, *Bombshell*, pp. 194–200; Aldrich, 'British Intelligence', pp. 331 ff.; Tchikov, *Comment Staline*, pp. 213, 220–21.

67 See Close, *Half-Life*, pp. 301–2, 309 and fns.11 and 12, pp. 360–61. By 1950 blueprints of the Canadian reactor were in the USSR and Pontecorvo is the prime suspect for passing these to the KGB (Close, p. 309). Andrew and Gordievsky, *KGB*, p. 327; Albright and Kunstel, *Bombshell*, pp. 133–4; Andrew and Mitrokhin, *Mitrokhin Archive*, p. 194.

68 Tchikov, *Comment Staline*, pp. 216–24; Andrew and Mitrokhin, *Mitrokhin Archive*, p. 194; Primakov, *Ocherki*, pp. 206–7 (Sokolov's memories of the exfiltration are in the official SVR history); Dolgopolov, *Legendarnye*, p. 319; Gromushkin, *Razvedka*, p. 95 (Gromushkin was awarded the Order of the Red Star for his forgery work for the Cohens and others); Pavlov, *Operatsiya*, p. 140.

69 Antonov, *Konon Molody*, pp. 65–8; Pavlov, *Operatsiya*, pp. 110, 141; Tchikov, *Comment Staline*, pp. 225–36; Chikov, *Sekretniye*, pp. 9–31; Drozdhov, *Nuzhnaya*, pp. 47, 83.

70 Tchikov, *Comment Staline*, pp. 237, 240–49; Antonov, *Konon Molody*, pp. 84–6. Wolf-whistle: Elwell note of Mrs Search interview, 20 December 1961: KV 2/4455. The Cohens' first radio was found buried in plastic in the back garden of 45 Cranley Drive only in 1980: see BBC *Southern Eye* documentary, 1993.

71 Chikov, *Sekretniye*, pp. 29–31, 43–4, 55, 71.

72 Tchikov, *Comment Staline*, pp. 237, 240–49; Antonov, *Konon Molody*, pp. 84–6. Andrew and Mitrokhin, *Mitrokhin Archive*, p. 533. Kolosov and Molody, *Mertvy sezon*, pp. 31–2.

73 Bulloch and Miller, *Spy Ring*, pp. 185–6; Andrew and Mitrokhin, *Mitrokhin Archive*, p. 535; Burke, *Spy Who Came in from the Co-op*, pp. 160–61.

74 Interview, BBC *Southern Eye* documentary, 1993.

75 Andrew and Mitrokhin, *Mitrokhin Archive*, p. 535; Burke, *Spy*, pp. 160–61; Carter, *Chemical and Biological Defence*, pp. 72–3, 75–84; Tucker, *War of Nerves*, pp. 146–7, 154–5, 158–9. 'Soviet Chemical Warfare', September 1960: DEFE 44/204. Information from Nigel Chapman, son of Norman Chapman, Professor of Chemistry at Hull University 1956–81, who worked at Cambridge University during the Second World War on chemical weapons, and afterwards at and with Porton Down, serving on various high-level committees. Tuorinsky, *Medical Aspects*, p. 157. Carter and Balmer, 'Chemical and Biological Warfare', p. 303.

76 Chikov, *Sekretniye*, pp. 84–6; Tchikov, *Comment Staline*, pp. 248–50; Kouzminov, *Biological Espionage*, p. 25.

77 Kouzminov, *Biological Espionage*, pp. 25–6. 'Russia knew West's germ warfare secrets', *Daily Telegraph*, 12 February 2005, https://www.telegraph.co.uk/news/uknews/1483350/Russians-knew-Wests-germ-warfare-secrets.html (accessed 1 June 2019). Molody's account of his Porton Down espionage work in his memoirs is typically misleading and tendentious but underscores the importance the USSR placed on this activity: Lonsdale, *Spy*, pp. 101–3. Primakov, *Ocherki*, p. 179. Antonov, *Konon Molody*, pp. 100–102. Kolosov and Molody, *Mertvy sezon*, pp. 114–15. 'Soviet Biological Warfare', para. 7: DEFE 44/205; 'Soviet Chemical Warfare', para. 13: DEFE 44/204; 'Soviet Bloc War Potential, 1969–73', JIC report, 17 January 1969, p. 53: CAB 186/1.

78 Primakov, *Ocherki*, p. 178; Antonov, *Konon Molody*, pp. 100–101; Antonov, *Vokopakh*, p. 119; Kolosov and Molody, *Mertvy sezon*, pp. 126–9; Lonsdale, *Spy*, pp. 94, 110–101, 117; Evseev, *Moya professiya*, pp. 113–24. On death of Bacon see Schmidt, *Secret Science*, pp. 280–83.

79 Tchikov, *Comment Staline*, pp. 249–51. No further information about Chambers, Elliott and Baron and their identities is known.

80 Lewis, *Spycatcher*, pp. 136–7; Houghton, *Operation Portland*, pp. 154–5.

81 BBC *Southern Eye* documentary, 1983; Lewis, *Spycatcher*, pp. 136–7; Houghton, *Operation Portland*, pp. 157–60.

82 Houghton, *Operation Portland*, pp. 55–58, 71, 87–9, 92–4; *Spectator*, 15 July 1972.

83 Interview, West. West and Tsarev, *Crown Jewels*, pp. 256–71, summarises the KGB 'Shah' files to which access was allowed and is the main basis of this chapter.

84 West and Tsarev, *Crown Jewels*, pp. 256–8.

85 Feklisov had controlled the Rosenberg network in New York during the Second World War and later was the case officer for Klaus Fuchs in London from 1947 to 1949.

86 West and Tsarev, *Crown Jewels*, pp. 259–63; Feklisov, *The Man behind the Rosenbergs*, p. 64.

87 West and Tsarev, *Crown Jewels*, pp. 266–8; Andrew and Mitrokhin, *Mitrokhin Archive*, p. 520.

88 A version of this was almost certainly the manuscript document found at Gee's house after her arrest.

89 Primakov, *Ocherki*, p. 178.

Epilogue

1 Tchikov, *Comment Staline*, pp. 329–30; Dolgopolov, *Legendarnye*, pp. 313, 323.

2 Lonsdale, *Spy*, pp. 120–21; Kolosov and Molody, *Mertvy sezon*, pp. 133–4; Dozhdalev interview in 1993 BBC *Southern Eye* documentary (in which he said Molody did alert the Centre to possible surveillance but Moscow did not order all operations of his network to cease); Tchikov, *Comment Staline*, pp. 312–13; NATO country source.

3 Lyubimov interview.

4 West and Tsarev, *Crown Jewels*, pp. 257–62, 268, 271; Primakov, *Ocherki istorii*, p. 178; Prince interview. See also Houghton, 'Notes for Counsel', attached to Hollis to Hunt, 11 May 1951: CAB 301/251.

5 Parry, 'History', in the section headed 'The Nuclear Revolution, Sonar 2001'; Parry interview.

6 An Introduction to the Admiralty Underwater Weapons Establishment Portland, 1963, pp. 1–2, 8–9: ADM 302/147; Scientific and Technical Progress Report 1961, AUWE Portland, pp. 1–15, 19–22: ADM 302/141; Weapon Control System for Nuclear Submarines, HMUDE pamphlet no. 653, October 1959, pp. 1–2: ADM 259/532;

Introduction to Sonar Type 2001, 1963: ADM 302/327; Introduction to ASDIC Type 184, AUWE, December 1961: ADM 302/198.

7 Igor Vodyanoy, 'Acoustic Research in Russia – a review', 16 August 1998: fas.org/nuke/guide/Russia/industry/docs/ivnews41.htm (courtesy of Nathan Prince) (accessed 25 August 2019).

8 Interview with Nairn (a sonar expert who worked at Portland from 1965 to 1998).

9 It seems the active component of 'Sector' was especially valuable, and of interest, to the Russians. This consisted of thirteen different transmitting modes, used in various combinations, e.g. high or low frequency, long or short pulse duration, direction or arcs of transmission. Russian concern about the effectiveness of Type 2001 active sonar probably helps to explain why they went to considerable lengths in the 1960s and early 1970s to develop 'rubber' (anechoic) tiles which were fixed to the outer hull of submarines to reduce the returning echo. Source: interview with John Edgell.

10 Polmar and Whitman, *Hunters and Killers*, pp. 141–2; Hennessy and Jinks, *Silent Deep*, pp. 320–21; Polmar and Moore, *Cold War Submarines*, pp. 77, 136–42, 156–8. Prince, Cust, Parry and Edgell interviews (the last three all commanded Royal Navy nuclear submarines during the Cold War; and Edgell was Director of Naval Intelligence Collection, 2000–03).

11 Polmar and Whitman, *Hunters and Killers*, pp. 141–2; 'Increasing Capability of the Soviet Maritime Forces', JIC report JIC(75)12, 20 June 1975, p. 3, para. 5: CAB 186/20. This report began by underlining that the 'magnitude of the improvements [in Soviet maritime capability] since 1970 has been unparalleled in any previous five-year period'. Some of these improvements probably owed something to the treachery of the Walker naval spy ring in the USA rather than the legacy of the secrets betrayed by Houghton and Gee. Interviews, Cust and Edgell.

12 Dozhdalev quote in BBC *Southern Eye* 1993 documentary; Antonov, *Vokopakh*, p. 120; Vitkovsky, *Poedinok spetssluzhb*, pp. 332–3.

13 Hennessy and Jinks, *Silent Deep*, pp. 548–50; Sulick, *American Spies*, pp. 93–107.

14 Parry, 'History'; Preston, 'Submarine Design', p. 175; 'Soviet Bloc War Potential, 1969–73', JIC report, ibid.; 'Soviet and East European General Purpose Forces', CIA National Intelligence Estimate no. 11-14-19, 4 December 1969, p. 23 (declassified 2017); 'Soviet Antisubmarine Warfare: Current Capabilities and Priorities', CIA Intelligence Report, September 1972, pp. 22–6 (declassified 2017).

The US navy 'Dimus' sonar, BQQ-2, had an even more powerful passive sonar than the 2001, housed in a sphere of hydrophones so it could electronically search in three directions: Polmar and Whitman, *Hunters and Killers*, pp. 91–2; Polmar and Moore, *Cold War Submarines*, p. 148; interviews, Cust and Edgell.

15 Antonov, *Konon Molody*, p. 172.

16 Rimington, *Open Secret*, pp. 118–19.

17 Whyte papers; Stell interviews.

18 Elwell, *An Autobiography*, pp. 97–102, 106–7, 112–13; Elwell obituary, *Daily Telegraph*, 22 January 2008 at https://www.telegraph .co.uk/news/obituaries/1576192/Charles-Elwell.html; Selina Elwell interviews.

19 Vol. 5 of Warren Commission, pp. 1–33: https://www.govinfo.gov /content/pkg/GPO-WARRENCOMMISSIONHEARINGS-5/pdf /GPO-WARRENCOMMISSIONHEARINGS-5.pdf ; Powers, *Broken*, p. 254; Belmont, 'As I recall it', chapter 49.

20 Evans, interview; Andrew and Mitrokhin, *Mitrokhin*, pp. 55–88, 212–29, 248–64, 328, 356–9; NATO source, interview; Vavilova, interview; Kroger and Lonsdale, *Pisma*.

21 Urban, *The Skripal Files*, pp. 173–84, 243–4; Corera, *Russians Among Us*, *passim*; Vavilova, interview.

22 Interview, Sipher; Corera, *Russians Among Us*, pp. 7–8.

23 Interview Abbott.

Afterword

1 Interview with Linda Horne (Houghton's granddaughter), 27 September 2020.

2 Various documents from FBI files on the Cohens, Jan and early Feb 1961, released to the author under FOIA, November 2020.

3 http://svr.gov.ru/history/person.htm (accessed Feb 2021)

Index

Picture credits

The author and publisher are grateful to the following for permission to reproduce photographs:

p.1 (top) Alamy/Mirrorpix; (bottom) author supplied c/o Stuart Morris

p.2 (top) The National Archives, Kew/Crown Copyright 2020; (bottom) Alamy/Mirrorpix

p.3 (top and middle) The National Archives, Kew/Crown Copyright 2020; (bottom) author supplied c/o Charles Elwell

p.4 (top and bottom) author supplied c/o Jocelyn Stell

p.5 (top and middle) author supplied c/o Charles Elwell; (bottom) author supplied c/o Selina Elwell

p.6 (top) author supplied c/o Vladimir Antonov; (bottom) Getty/Stringer

p.7 (top) GCHQ; (middle) Alamy/Mirrorpix; (bottom) author supplied c/o Vladimir Antonov

p.8 (top) Mirrorpix/Norman Lucas; (bottom) US National Archives

p.9 (top) author supplied c/o Gordon Lonsdale, *Spy* (London, Neville Spearman: 1965); (middle) Getty/Stringer; (bottom) Getty/Popperfoto

p.10 (top and bottom) US National Archives

p.11 (top left and right) Getty/Popperfoto; (middle and bottom) The National Archives, Kew/Crown Copyright 2020

p.12 (top) The National Archives, Kew/Crown Copyright 2020; (middle) author supplied c/o Vladimir Antonov

p.13 (top) Alamy/Mirrorpix; (bottom) Alamy/Keystone Press

p.14 (top) Alamy/Mirrorpix; (bottom) Imperial War Museum

p.15 (top) author supplied c/o Gordon Lonsdale, *Spy* (London, Neville Spearman: 1965)

p.16 (top) author supplied c/o SVR Press Bureau; (middle and bottom) author supplied c/o Andrey Vedyaev

p.159 The National Archives, Kew/Crown Copyright 2020